Emerging ICT Technologies and Cybersecurity

Kutub Thakur · Al-Sakib Khan Pathan · Sadia Ismat

Emerging ICT Technologies and Cybersecurity

From AI and ML to Other Futuristic Technologies

 Springer

Kutub Thakur
Department of Professional Security
Studies
New Jersey City University
Jersey City, NJ, USA

Al-Sakib Khan Pathan
Department of Computer Science
and Engineering
United International University
Dhaka, Bangladesh

Sadia Ismat
Department of Professional Security
Studies
New Jersey City University
Jersey City, NJ, USA

ISBN 978-3-031-27767-2   ISBN 978-3-031-27765-8   (eBook)
https://doi.org/10.1007/978-3-031-27765-8

This Springer imprint is published by the registered company Springer Nature Switzerland AG
The registered company address is: Gewerbestrasse 11, 6330 Cham, Switzerland

*To my wife Nawshin Thakur and my children
Ezyan Thakur and Eyana Thakur*

—Kutub Thakur

*To my wife Labiba Mahmud and two little
daughters: Rumaysa Sakib and Rufaida Sakib*

—Al-Sakib Khan Pathan

*I would like to dedicate this book to my
husband Zahir Ahmed for his continuous
support and encouragement and to our two
sons Zarif Ismat Ahmed and Zafri Ismat
Ahmed for inspiring me everyday*

—Sadia Ismat

Preface

This book begins with the topics related to advanced information and communications technology (ICT), then artificial intelligence (AI)/machine learning (ML) technologies (that are more or less known and some unknown or are still being explored), then talks about various emerging technologies like tactile virtual reality issues, 6G, 4D printing, edge computing, then goes on to introducing futuristic technologies like Space Elevator, Rotating Skyhook, Mind Uploading, Human 2.0 initiative (which is somewhat controversial), etc., and then, it links all these issues with cyber and cybersecurity issues. Hence, the main focus is not necessarily cybersecurity but as in the discussion, we have brought here various emerging and futuristic technologies and concepts—we have tried to assess what their impact could be on cybersecurity as all these would expand the attack surface for cyberspace.

If indeed a human's mind can be influenced or programmed with AI via cyber channel (remotely), that is even more dangerous than what we know in today's world. Is it really possible? But, some are thinking in that direction! The fact of the matter is that many of these ideas may not ever become a reality, but these are available or are coined in the technical and scientific domains. How the humankind would be with technology all around them in the coming days or whether all these are at all beneficial for the humankind (for their existence) is analyzed and commented on while describing the concepts, current status, and advancements. As far as we have investigated the current books available in the field, no book has compiled all these in one single volume so far. Hence, we are hopeful that this would be something of interest as a textbook for the graduate and undergraduate students and also as a reference book for the researchers. Especially, it needs to be mentioned that in some courses, these topics are taught separately at the graduate/undergraduate levels or for research students, but we thought if a book shape is given to all these, that could be really helpful as a premier reference source.

We sincerely thank the Almighty Allah to allow us the time, energy, and patience to complete this project. Special thanks to **Annelies Kersbergen**, who has helped us throughout the period of the book's proposal and approval stages.

Jersey City, USA   Dr. Kutub Thakur
                                                                    kthakur@njcu.edu

Dhaka, Bangladesh   Dr. Al-Sakib Khan Pathan

Jersey City, USA   Dr. Sadia Ismat
                                                                    sismat@njcu.edu

Contents

About the Authors

Kutub Thakur is Director of NJCU Center for Cyber-security and Assistant Professor and Director of Cyber-security Program at New Jersey City University. He worked for various private and public entities such as United Nations, New York University, Lehman Brothers, Barclays Capital, ConEdison, City University of New York, and Metropolitan Transport Authority. He received his Ph.D. in Computer Science with special-ization in cybersecurity from the Pace University, New York, M.S. in Engineering Electrical and Computer Control systems from University of Wisconsin, and B.S. and A.A.S. in Computer Systems Technology from the City University of New York (CUNY). He reviewed for many prestigious journals and published several papers in reputable journals and conferences. His research interests include digital forensics, network security, machine learning, IoT security, privacy, and user behavior. Dr. Thakur is currently serving/served as Program Chair for many conferences and work-shop. He is also currently supervising (also, supervised) many graduate and doctoral students for their theses, proposals, and dissertations in the field of cybersecurity.

Al-Sakib Khan Pathan is Professor at Computer Science and Engineering department, United International University (UIU), Bangladesh. He is also serving as Ph.D. Co-supervisor (external) at Computer Sciences Department, University Ferhat Abbas Setif 1, Algeria. He received Ph.D. degree in Computer Engineering in 2009 from Kyung Hee University, South Korea, and B.Sc. degree in Computer Science and Information Technology from Islamic University of Technology (IUT), Bangladesh, in 2003. In his academic career so far, he worked as Faculty Member in various capacities in various institutions like the CSE Department of Independent University, Bangladesh (IUB), during 2020–2021; Southeast University, Bangladesh, during 2015–2020; Computer Science department, International Islamic University Malaysia (IIUM), Malaysia, during 2010–2015; BRACU, Bangladesh, during 2009–2010; and NSU, Bangladesh, during 2004–2005. He served as Guest Professor at the Department of Technical and Vocational Education, Islamic University of Technology, Bangladesh, in 2018. He also worked as Researcher at Networking Lab, Kyung Hee University, South Korea, from September 2005 to August 2009 where he completed his M.S. leading to Ph.D. His research interests include wireless sensor networks, network security, cloud computing, and e-services technologies. Currently he is also working on some multidisciplinary issues. He is a recipient of several awards/best paper awards and has several notable publications in these areas. So far, he has delivered 32 keynotes and invited speeches at various international conferences and events. He was named on the list of Top 2% Scientists of the World, 2019 and 2020 by Stanford University, USA, in 2020 and 2021. He has served as General Chair, Organizing Committee Member, and Technical Program Committee (TPC) Member in numerous top-ranked international conferences/workshops like INFOCOM, CCGRID, GLOBECOM, ICC, LCN, GreenCom, AINA, WCNC, HPCS, ICA3PP, IWCMC, VTC, HPCC, SGIoT, etc. He was awarded the IEEE Outstanding Leadership Award for his role in IEEE GreenCom'13 conference and IEEE Outstanding Service Awards twice in recognition and appreciation of the service and outstanding contributions to the IEEE IRI'20 and IRI'21. He is currently serving as Editor-in-Chief of *International Journal of Computers and Applications* and *Journal of Cyber Security Technology*, Taylor & Francis, UK; Editor of *Ad Hoc and Sensor Wireless Networks*, Old City Publishing, *International Journal of Sensor Networks*, Inderscience Publishers, and *Malaysian Journal of Computer Science*; Associate Editor of *Connection Science*, Taylor & Francis, UK, *International Journal of Computational Science and Engineering*, and Inderscience; Guest Editor of many special issues of top-ranked journals; and Editor/Author of 32 books. One of his books has been included twice in Intel Corporation's Recommended Reading List for Developers, 2nd half 2013 and 1st half of 2014; 3 books were included in IEEE Communications Society's (IEEE ComSoc) Best Readings in Communications and Information Systems Security, 2013; and several other books were indexed with all the titles (chapters) in Elsevier's acclaimed abstract and citation database, Scopus and in Web of Science (WoS), Book Citation Index, Clarivate Analytics; at least one has been approved as a textbook at NJCU, USA, in 2020; one is among the top used resources on SpringerLink in 2020 for UN's Sustainable Development Goal 7 (SDG7)—Affordable and Clean Energy; and one book has been translated to simplified Chinese language from English version. Also, 2 of his journal papers and 1 conference paper were included under different categories in IEEE Communications Society's (IEEE ComSoc) Best Readings Topics on Communications and Information Systems Security, 2013. He also serves as Referee of many prestigious journals. He received some awards for his reviewing activities like: one of the most active reviewers of IAJIT several times and Elsevier Outstanding Reviewer for Computer Networks, Ad Hoc Networks, FGCS, and JNCA in multiple years. He is Senior Member of the Institute of Electrical and Electronics Engineers (IEEE), USA.

Sadia Ismat has been serving government agencies for the last two decades. She has held multiple executive positions. Currently, she is working as Chief Information Security Officer (CISO) for the Department of Finance, New York City's largest financial municipal agency which has a collective revenue annually of 65 billion dollars. Dr. Sadia has worked for various state and city agencies, providing consulting services to government and private firms throughout her career. She is regular Contributor and Speaker in Cybersecurity Conferences discussing Cybersecurity and Women in Cybersecurity. Dr. Ismat published research works encompassing federal compliance and biometric technology. She also works as Adjunct Professor to various city and private universities.

Chapter 1
An Overview of ICT Technology Advancement

Introduction

Information and Communication Technology, precisely referred to as ICT, is a wider umbrella term that covers different types of technologies—both hardware and software technologies—associated with the communication systems, techniques, transmission media, software programs, protocols, sensors, and standards.

The most common examples of ICT technologies include all modern automated systems powered by modern computers, storage devices, transmission media, software platforms, and other related elements including web-based data[1] and voice communication systems, integrated and embedded systems, Internet of Things (IoT), Cyber-Physical Systems (CPS), Internet, and many others.

Cyber-Physical Systems (CPS), Network of Networks (i.e., Internet), and Internet of Things (IoT) are a few crucial domains of ICT that are growing very fast and changing the landscapes of all types of businesses, processes, infrastructures, industries, and societies across the globe tremendously. CPS is a wide range of modern systems that have revolutionized numerous industries such as energy grids, civic systems, healthcare, industrial automation, self-driven automobiles, robotics, home automation and many others. A CPS system is a computer-based system whose core mechanism of operation and function is monitored and controlled by computer programs, commonly referred to as software algorithms or applications and integrated through communication networks for effective coordination and communication among the ICT elements [1]. Substantial advancements in CPS and other ICT technologies have left highly desirable and unprecedented positive impact on all industries, infrastructures, and societies worldwide.

Driven by the challenges of system security, reliability, speed, performance, complexity, process automation, market competitiveness, societal development, industrial growth, and others, the demand for faster advancements in all the domains of ICT is very high today. Ever-demanding end-users have initiated huge developments and enhancements to achieve various desirable factors such as [2]:

- Faster, reliable, and efficient communication
- Consistency, continuity, and complex networking solutions
- Competitive and intelligent ICT systems
- Big data analysis and business intelligence
- Reliable process, office, home, and industrial automation
- Building automated and connected societies and governments.

Indeed, the technological developments shaped various sectors in ways that we could not even imagine just a few decades ago. ICT is the main driving force for all these, especially when it comes to linking people in various parts of the globe and for connectivity that also opened new areas of trade, exchange, and interactions. The technological growth was so rapid during the past couple of decades that it transformed each and every thing in the world. In fact, ICT still has a tremendous potential to transform both societies and economies in different ways like for instance [3, 4]:

- Reduction in transaction and information cost
- Improvement in access to basic services and education
- Enhancement in the efficiency of business processes
- Improvement in worker productivity and efficiency

---

[1] In this book, the term '*data*' has been used both as singular and plural, depending on the context.

- Promotion of innovative approaches to problems
- Reduction in operations and maintenance cost
- Expansion in domains of industries and services
- Globalization of industries and services
- Healthcare and biotech engineering revolution
- Enhanced information sharing for better interconnection among users and businesses.

The majority of these impacts are achieved through the advancements in the core technologies that work behind the ICT services and products. Let us have an overview of ICT advanced technologies that leave highly desirable impact on our social and personal lives as well as the processes of businesses across the globe.

An Overview of ICT Advanced Technologies

Technology powered by the modern computing, transmission media, networking techniques, software applications, and electronic storages is commonly referred to as an Advanced ICT technology. The emergence of modern ICT technology can be traced back in early 40s when the Word War II (WWII) was going on. The first commercial computer developed in 1951 was named as UNIVAC I. We can say that the start of the modern ICT started from that point [5]. The image of UNIVAC I computer is shown in the following Fig. 1.1.

After the invention of UNIVAC-I computer, the majority of the research and development in the field of ICT was concentrated in the defense industry to create advanced technologies focused on military-use to develop competitive-edge and dominance in the warfare domain. The term Information Technology (IT) was first time coined in around 1970, which led to the development of new term, ICT later in the field of modern technologies [6].

Fig. 1.1 UNIVAC I computer (Flickr)

In the later sections of this chapter, an overview of the history of ICT will be presented in the form of generations of different technologies associated with the ICT field.

Main Areas of ICT Technologies

The progress and development of ICT can be categorized into two major areas such as:

- Hardware Technologies
- Software Technologies.

The advancement of any technology related to IT hardware and software is directly associated with the advancement of ICT. The details of hardware and software categories are explained in the following sections separately.

Hardware Technologies

Hardware technologies used in ICT field include the capabilities of different electronic devices or materials for processing, storing, transporting, inputting, and outputting data or information in the form of signals. These capabilities of hardware materials or equipment are powered by the supported software protocols, techniques, and algorithms. The hardware technologies normally work on the basis of the properties and capabilities of hardware. The most common capabilities and properties of a hardware include [7]:

- Physically tangible IT components, i.e., that can be touched and felt
- Capable of handling data processing, transmission, and storage
- Ability to run software instructions
- Hardware components are manufactured through certain processes
- Not affected by the computer viruses and other malware software.

The above-mentioned properties of IT hardware components can materialize a range of hardware technologies or IT processes used in ICT technologies. The main area of hardware technologies can be further divided into the following categories:

- Data processing hardware
- Data input hardware
- Data output hardware
- Data transmission hardware
- Data storage hardware.

Data Processing Hardware

Data processing is an electronic technology powered by the associated software program to run different types of operational processes on the electronic signals to generate different results based on those operations such as NOR, NAND, OR, addition, multiplication, and others. Different electronic components are used to perform those operations within an electronic hardware-based structure known as the integrated circuits, boards, and other components. The examples of data processing hardware include the following:

- Microprocessors
- Microcontrollers
- Central Processing Units
- Graphics cards
- VGA (Video Graphics Array) cards
- Network cards
- Sound cards.

The technologies that substantialize the working capabilities of these ICT processing devices is known as the IT data processing technologies. The data processing can be done in two major categories—analog signal processing and digital signal processing. Both types of signal processing in the data processing elements use numerous types of algorithms, techniques, and operations that are referred to as the basic parts of electronic processing technologies. Those technologies include a range of signal processing mechanisms and algorithms in both manual and digital signal processing in our modern IT hardware elements. A few of those most common data processing technologies and techniques used in the hardware components include [8, 9]:

- Addition, multiplication, and delay functions commonly referred to as processing operations
- Forward feeding and backward feeding processes
- Signal sampling process
- Signal quantization technique
- VLSI (Very Large-Scale Integration) signal processing
- Signal analyzing and synthesizing
- Analog to digital A/D conversion
- Digital to analog D/A conversion
- Superimposition of signals
- Modifying and clipping of signal waves
- Aliasing of signals
- Filtering of signals
- Formatting of signals
- Range of algorithms for processing signals include Fourier transformation, Linear Time Variant Systems, convolution process, Windowing processes, and others.

Data processing hardware can be classified into two major categories—standalone computer processing hardware and network element data processing hardware. The standalone processing devices can be like CPU/GPU (Graphics Processing Unit), video/sound card, and so on. The network-based hardware used for data processing include routers, switches, hubs, access points, gateways, firewalls, bridges, and other elements connected within a network [10].

Data Input Hardware

There are a large number of data input hardware that have achieved new advancements in their respective technologies. The data input hardware can be either simple data input devices that feed data manually to the computer processing units or automated signals through range of sensors in the modern IT processing applications. The common types of manual data input hardware for a computer can be a mouse, keyboard, camera, joystick, and others. The sensor-based hardware elements used for inputting the data into the data processing units include:

- Temperature sensor
- Barometers
- Microphone
- Biometrics scanners
- Bar code readers
- Cameras
- Oscillators
- Light Emitting Diodes (LED).

All these signal producing sensors used for inputting the data signals use different technologies to input the data to the processing devices. A few of the input techniques commonly used in the modern processing devices include [11]:

- Input transducing technique
- Point-and-click technique
- Pinch-to-zoom technique in touch screen input commonly known as interaction technique
- Different user interfaces and ports
- Status sensed technique
- Device acquisition time sensor technique
- Dimension sensing technique
- Property sensed technique.

Data Output Hardware

The output hardware is a type of ICT hardware that is used to display, present, or express the output data or information, which is processed by the processing hardware. The most common IT output hardware include computer monitors, plotters, printers, projectors, GPS (Global Positioning System) devices, speech synthesizers, headphones, speakers, and others. These devices can be connected to the processing hardware in standalone as well as networked environments to present the output information in the designated formats that are easily understandable to human beings or any other desired target [12].

In terms of a networked system, an output hardware is the electronic device used to send data from one machine to another. The main examples of output devices in a network ecosystem include RFID (Radio Frequency Identification) devices, IR (Infrared) gadgets, network cards, modems, solenoids, and many others. These devices normally work as both input as well as output devices in the networking environment [13].

The main technologies used in the output computing devices include different data formats such as JPEG (Joint Photographic Experts Group), MP4, sine waves, digital signaling technologies, text printing, image printing, sonic waves, light signals, and so on.

Data Transmission Hardware

Data transmission hardware is a range of equipment that consists of transmission media, connectors, and physical interfaces that connect data sender and data receiver devices. The role of transmission media and associated equipment is very critical in the modern ICT systems. The most common types of transmission hardware, media, equipment include:

- Ethernet cables
- Fiber cables
- Twisted pair cables
- Coaxial cables
- Serial cables
- Parallel interfaces.

The most common types of data transmission modes can be categorized into the following classes [14]:

- Half-duplex transmission
- Full-duplex transmission
- Simplex transmission
- Serial communication
- Parallel data transmission.

The technologies governing the transmission of the data between two end devices may include the transmission protocols, development of new transmission materials, refinement of transmission techniques, development of modern air transmission technologies and new standards for faster transmission of data through fiber, air interface, and copper materials. The most common examples of modern technologies used in the data transmission include:

- Latest wireless technologies such as 4G, 5G, cellular, Wi-Fi, and others
- Fiber transmission through SDH (Synchronous Digital Hierarchy), DWDM (Dense Wavelength Division Multiplexing), and others
- Satellite communication protocols
- Serial communication standards and interfaces
- Parallel communication protocols and interfaces.

Data Storage Hardware

Information Technology hardware used for storing the data in different forms and formats for the future use is known as the storage hardware. The most common types of data storage hardware used in the modern ICT systems include [15]:

- Hard Disk Drives (HDD)
- Solid State Drives (SSD)
- Magnetic tapes
- Optical cartridges
- Read Only Memory (ROM)
- Random Access Memory (RAM)
- CD (Compact Disc), DVD ("Digital Video Disc" or "Digital Versatile Disc"), Blu-Ray Discs
- USB (Universal Serial Bus) Flash memory.

The storage devices are usually controlled by the electronic and magnetic technologies that govern the storage process of data on those hardware devices. The most common examples include NAND flash electronic transistors, magnetic codes, flip-flops, and others.

Software Technologies

The major software technologies that are growing and emerging in the domain of information and communication technology can be divided in different categories such as operating systems (OSs), protocols, programming languages and platforms, and development methodologies. The impact of the software technologies on the overall domain of ICT is huge because they play very vital role in the materialization of the modern IT ideas, applications, devices, equipment, and industries. Without the power of software technologies, it is not possible to build the most modern and cutting-edge inventions in all fields of industries such as healthcare, automobile, industrial automation, smart cities, home/office automation, online web ecosystems, telecommunication, transportation, and others. It is the software that brings a device to life by interacting with the other devices and people. The software programs in different categories such as firmware, OS, and apps allow machines to load and run on a range of hardware to operate and control the functionalities of those hardware devices.

Firmware

A firmware is a kind of software code that is developed for a small device-specific hardware to provide low-level control over the operations of that particular electronic device. A firmware can be further classified into two classes such as updateable and fixed firmware. A fixed firmware is loaded once on the embedded device at the time of building an equipment or electronic module. That firmware is not updated again and again and works for a very short and specific purpose to provide smaller control on a particular action of the device. The upgradable firmware can be regularly upgraded to resolve bugs and other software issues through patches. The enhanced version can also be added to replace the existing firmware for providing additional features, capabilities, and security. A flashing speed controlling firmware in a camera is loaded on a chip as shown in the Fig. 1.2.

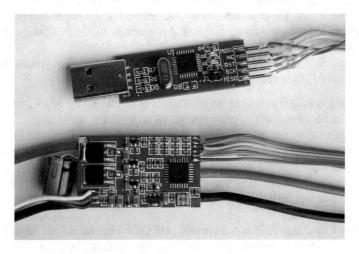

Fig. 1.2 Flashing speed controlling firmware installed (Flickr)

The main features of a firmware are mentioned below [16]:

- Provides hardware abstraction services to higher level control software like full-fledged OS
- Acts as the basic operating systems for smaller/simpler devices with very fundamental functions
- A firmware is normally copied/installed/held on non-volatile memories such as flash, ROM, EEPROM (Electrically Erasable Programmable ROM), and EPROM.
- The term "firmware" was first coined in 1967 by Ascher Opler in Datamation computer journal
- A few examples of firmware include BIOS (Basic Input/Output System), POST, RTAS, home/office appliance controllers, industrial machine controllers, and others
- Firmware is written in low-level languages such as machine/assembly language to build a short code for specific function/activity to achieve.

Operating Systems (OSs)

Operating system, precisely referred to as OS, is a larger software program that is designed to provide high-level control over a complex device to perform a wide range of functions, tasks, and activities under a hardware environment, especially computer-powered devices, equipment, and tools. The most common capabilities of an operating system include management of computer hardware, running common services, allocation of computer resources, interaction with software applications and utilities, and others. The coordination among different applications and processes is also performed by the operating systems. An interface of Linux Ubuntu operating system is shown in Fig. 1.3.

The main features, attributes, and capabilities of an operating system (OS) are mentioned in the following list [17]:

- Operating system is a general-purpose software to manage hardware and software resources on a computer or computer-powered complex machines.
- In a layered structure, an operating system works above the firmware that resides on the non-volatile storage on computer, to establish a high-level control over the hardware.
- The interaction among different machines, network elements, external hardware devices, internal software applications, and OS services is also performed by an operating system.
- Modern operating systems can provide both GUI (Graphical User Interface) and command prompt interfaces to interact with the operating systems.
- Operating systems are continuously improved and patched to counter any emerging security and other threats.
- Operating Systems can be classified in a range of categories as mentioned in the following list:

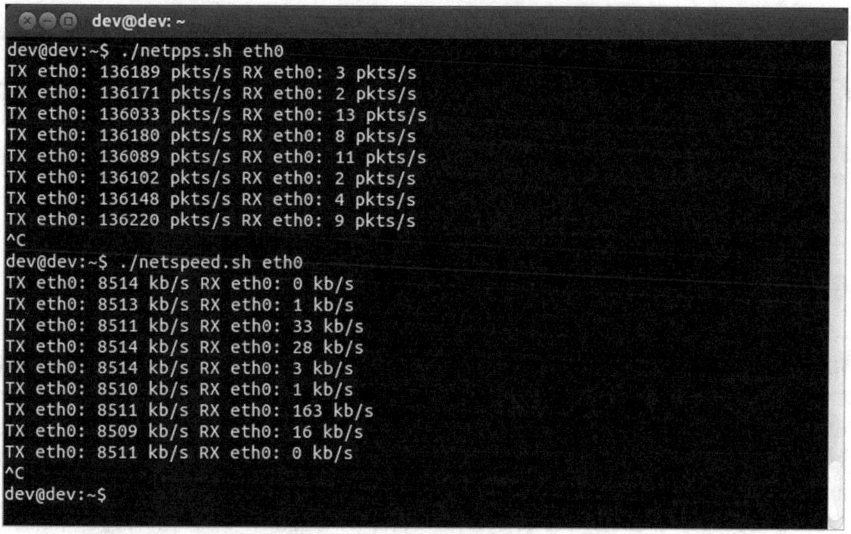

```
⊗ ⊝ ⊡    dev@dev: ~
dev@dev:~$ ./netpps.sh eth0
TX eth0: 136189 pkts/s RX eth0: 3 pkts/s
TX eth0: 136171 pkts/s RX eth0: 2 pkts/s
TX eth0: 136033 pkts/s RX eth0: 13 pkts/s
TX eth0: 136180 pkts/s RX eth0: 8 pkts/s
TX eth0: 136089 pkts/s RX eth0: 11 pkts/s
TX eth0: 136102 pkts/s RX eth0: 2 pkts/s
TX eth0: 136148 pkts/s RX eth0: 4 pkts/s
TX eth0: 136220 pkts/s RX eth0: 9 pkts/s
^C
dev@dev:~$ ./netspeed.sh eth0
TX eth0: 8514 kb/s RX eth0: 0 kb/s
TX eth0: 8513 kb/s RX eth0: 1 kb/s
TX eth0: 8511 kb/s RX eth0: 33 kb/s
TX eth0: 8514 kb/s RX eth0: 28 kb/s
TX eth0: 8514 kb/s RX eth0: 3 kb/s
TX eth0: 8510 kb/s RX eth0: 1 kb/s
TX eth0: 8511 kb/s RX eth0: 163 kb/s
TX eth0: 8509 kb/s RX eth0: 16 kb/s
TX eth0: 8511 kb/s RX eth0: 0 kb/s
^C
dev@dev:~$
```

Fig. 1.3 Linux OS command interface (Flickr)

- Single- and multi-tasking operating systems
- Distributed operating systems
- Single- and multi-user operating systems
- Real-time operating systems
- Embedded operating systems
- Templated operating systems
- Library operating systems.

• The main components of an operating systems can be classified into the following categories:

- Kernel
- Networking
- User interface
- Security.

• The kernel of an operating system handles the following functions and processes:

- Interrupts
- Program execution
- Memory management
- Device derivers
- Disk access and file systems
- Modes
- Virtual memory
- Multitasking.

- The examples of operating systems include Windows, Linux, Unix, Ubuntu, iOS, Android, MacOS, MS DOS, TOS/360, DOS/360, and so on.
- An operating system normally resides on the Hard Disk Drives (HDD) of computers.
- Operating systems can be classified into four generations that are described later in this chapter.

IT Protocols

There is a significant parallel between two very important terms in the Information and Communication Technology (ICT) field—protocol and algorithm. Both of those terms are used in the field of ICT for establishing the modern communication among the network elements, databases, and software-based IT entities in a communication and information technology ecosystem.

IT protocol is a set of rules, syntax, communication synchronization, semantics, and communication error recoveries for establishing a smooth and predictable communication between two or more than two network elements. IT protocols are mostly referred to as communication protocols, which are based on standard seven layers of network protocols commonly referred to as Open System Interconnection (OSI) model. Any communication model can deal in predefined functionalities of one or more than one layers of the OSI model. Communication protocols use messages to exchange between two entities to establish communication based on predefined rules, syntax, and other necessities of smooth and effective communication [18].

Another very important domain of ICT technology is the programming language or algorithms that would handle the communication among the software-based resources/infrastructure through the exchange of data. Protocols handle the communication and algorithms handle the data flow in computation. Communication protocols use the flow of messages for accomplishing the designated tasks while the algorithms or software programs use data structure and exchange of data flow to achieve the desired purposes of the algorithm or a program in the modern field of programming or programmable computation.

The advancement in both communication protocols and programming languages or algorithms plays a very crucial role in the modern ICT field. Both of them are the fundamental drivers of the enhancement of technologies in the ICT field.

Programming Languages

A programming language is the library of the computational rules or vocabulary to instruct a machine/computer to perform a certain task automatically. Programming language has a special pattern to write code for computational instructions that is known as the syntax of the programming language. The other major component of a programming language is the semantics, which can also be considered as the meaning of the code for the machines to understand. Different programming languages have different syntaxes and semantics but the core purpose of all different types of computer programming languages is almost the same [19].

The main purpose of computer programming language is to build a software program that can implement, manage, initiate, and terminate the algorithm for performing an automated task through computation by exchanging the data among the concerned entities. The most common features, capabilities, and attributes of programming languages include [20]:

- Programming language allows you to interact with computers/machines.
- Computer programming language helps streamline the power of computing all endeavors made by human beings in the modern field of information technology.
- It allows building intelligence machines, software-defined-infrastructures, and automation of a range of tasks in various walks of life and society in modern life and businesses equally.
- Programming languages are the life to every process, product, and service in the modern life that heavily depends on the ICT technologies.
- Programming languages can be categorized into numerous classes such as:

  - High-level languages
  - Machine languages
  - Assembly languages
  - Scripting languages
  - Domain-specific languages
  - System languages
  - General-purpose languages
  - Esoteric languages
  - Visual languages.

- A few languages can be classified into two or more categories. For instance, C language is considered as a system as well as a high-level language; while Perl is referred to as scripting and high-level language.

Software Development Methodologies

The impact of software development methodologies on the growth and advancements of modern information technology is huge. A proper software development methodology can help developers, engineers, consultants, and IT pioneers to foresee the right need of idealistic demand of the technology market and plan the development in such a way that it achieves the desired targets in a very short span of time as well as under the competitive budgets. A good development methodology allows the developers and researchers sufficient space for modification without any wastage of notable volumes of resources to meet the pace of the latest technologies in the world.

A structured method for organizing the entire software development lifecycle such as planning, developing, coding, testing, debugging, deploying, supporting, and monitoring for suitable feedback of a particular shippable software product is known as the software development methodology [21].

There are numerous software development principles and methodologies most commonly used in the software development field such as:

• Waterfall Development Methodology
• Extreme Programming (XP)
• Agile Software Development (ASD) Methodology
• Lean Development Methodology
• Feature-Driven Development (FDD)
• Test-Driven Development (TDD)
• Scrum Software Development Methodology
• Prototype Model Development
• Rapid Application Development Model
• Dynamic System Model.

Agile Software Development (ASD) has become one of the most popular methodologies of software development in the modern era. ASD is based on the iterative method in which short iterations are used for the development. Numerous versions of Agile development are used to fit the custom development in a range of industries, companies, and teams. Different iterative methods of software development have decreased the project time period and budgets significantly. At the other end, these methodologies have increased the effectiveness, performance, change management, and profit margins significantly. Those desirable factors leave a great impact on the advancements of modern technologies, which are mostly driven by the cutting-edge software algorithms, techniques, and protocols [22].

Evolution of Information Technology

The evolution of information technology encompasses a wide range of technologies in different domains such as engineering hardware, software, data storage, transmission media, internetworking methodologies, data security, and many others. The starting point of the information point may not look very related to the modern types of IT components; but the fact is that the starting point of the IT evolution dates back to the development of an abacus in about 2400 BC for accomplishing a few mathematical operations physically and manually. This may be taken as the starting of the initial concept of binary system of communication, which is extensively used in the modern information technology spheres [23].

The evolution of information and communication technology (ICT) is based on different generations of technologies in the following areas:

- Computers Generations
- Operating System Generations
- Application Software Generations
- Programming language Generations
- Wireless Technologies Generations
- WWW (World Wide Web) Generations
- Storage Technology Evolution
- Software Development Generations.

Let us go through an overview of different generations of the IT areas that have contributed to the advancement of ICT technologies to the present level.

Computer Generations

Modern digital computer has evolved through numerous areas of engineering such as mechanical engineering, electromechanical engineering, electronic and modern digital engineering based on software engineering. Before we have a look into the modern electronic/digital computer generations, let us have a look into the earlier developments, especially the mechanical and electromechanical engineering developments that had paved the way for the advancements of the modern computer technologies [24].

- Abacus calculator—2700 BC
- Pascal calculator—1652 AD
- Stepped Reckoner—1694 AD
- Arithmometer—1820 AD
- Comptometer—1887 AD
- Difference Engine—1822 AD
- Analytical Engine—1834 AD
- Millionaire—1893 AD.

In some references, the above inventions in this field starting from the Pascal calculator through the first-generation inventions of computers are referred to as the zero generation machines as mentioned in the following topic [25].

Zero Generation (1642–1945)

This generation of computers is also known as the mechanical computer era. The most common examples of the computer machine development in this tenure are (see the above list) like the Pascal calculator, Leibniz calculator, Babbage's Difference Engine, and so on.

First Generation (1945–1954)

The first generation of the modern computers is also referred to as the "Vacuum Tube Computer Era". In this generation, the core technology used for processing, transmitting, and storing the data was vacuum tubes. The examples of first-generation computers include Colossus made by British Government, ENIAC project, and UNIVAC I computers.

Second Generation (1954–1963)

The second-generation computers were much faster, compact, and efficient as compared to the first-generation computers, which used vacuum tube technology. The second-generation computers are based on an advanced technology known as electronic transistor for switching, processing, and storing the electronic signals in this generation of computers. The main examples of 2nd generation computers include IBM 1400 Series and DEC PDP-I Series computers.

Third Generation (1963–1973)

The third-generation computers are classified based on the integrated circuit (IC) technology, which is much faster than individual transistor. Thus, more advanced, efficient, and capable computer machines were produced based on the integrated circuit technology for the third-generation computer era. The advent of the Large-Scale-Integration (LSI) technology also falls in this category that led to a very large integration of the transistor technology to increase the capacity and capabilities of the modern computing machines. The example of the computers pertaining to the 3rd generation include IBM PC, Intel 4004 chip, and others. The integration of over 6000 transistors were done in the Intel 8080 processor of this generation.

Fourth Generation (1973–1985)

The fourth-generation computers are characterized by the scale of integration of transistors. This generation of computers used the integration technology known as "Very Large-Scale Integration", precisely referred to as VLSI technology. The examples of computers of the 4th generation include MITS Altair 8800, IBM PC, and Apple computers. The number of transistors integrated into a fourth-generation computer reached to 1.2 million transistors in Intel processor 80,486 introduced in

1989. The number of transistor integration grew drastically in the later versions of the computers.

Fifth Generation (1985–Present)

The 5th generation computers are characterized by the advanced technologies supporting parallel processing, networking of the computers, and other capabilities. The integration of transistors in this generation of computers is referred to as "Ultra Large-Scale Integration", or, more precisely referred to as ULSI technology. In this generation, parallel processors (to share processing tasks) were deployed. Numerous other advancements in the storage, parallel communication, processing, and other technologies are also integrated in these computers. The main examples of this generation include all tablets, PCs (Personal Computer), mobile devices, and other computers commonly used in the present time.

Operating System Generations

Operating Systems are the software programs that manage computer software and hardware resources and offer basic services for the software applications running on the computers. The operating systems normally reside on the hard disk drive and initiated by the firmware of the computer embedded on the ROM [26].

The initial computers were designed for accomplishing simple computational functions and used the direct machine code created by the users or programmers. In early days, the users of computer used to be highly skilled and expert programmers to code certain functions to run. With the passage of time, operating systems took over the basic functioning of the computers and the users would need lesser skills and computer expertise to run the applications on the computer. The operating systems can be divided into different generations or OS eras as mentioned below [26, 27].

First Generation (1940–1950)

The first generation of operating system was the direct operating access on the computer to the programmers to develop a code in machine language and run on the computer. This tenure lasted for about a decade till early fifties. There is no particular recognized operating system to run the modern electronic computers, which would be used for simple mathematical calculations with the help of short codes created in machine language.

Second Generation (1955–1965)

The second-generation operating system is marked by the first specialized operating system developed by General Motors in early 1950s. This operating system was named as General Motors Operating Systems (GMOS) and was developed for the IBM computers. The main characteristics of this operating system included single stream batch processing and the use of punch cards for input jobs. This OS also required a technically skilled and professional operator to run tasks and transfer control to the operating system.

Third Generation (1965–1980)

This OS generation is characterized by the multiprogramming functionality in which multiple tasks would be performed through operating system. This era of operating system kick-started the use of minicomputers and paved the way for more advanced personal computers that fall under fourth-generation era of operating systems.

Fourth Generation (1980–Present)

The start of modern operating systems is referred to as the fourth-generation OS era. In this era, a range of modern operating systems were developed. The main features, characteristics, and examples of this generation of operating systems include:

- Most of the operating systems were driven by the personal computer systems.
- Microsoft Disk Operating System (MS DOS) was the first prominent operating system for personal computers.
- A series of Microsoft **Windows** operating systems are other major examples developed by Microsoft Corporation. All these operating systems supported graphical user interface (GUI), which led to pervasive use of personal computers among normal users without any programming skills and expertise.
- Macintosh Operating Systems, precisely known as Mac OS, was developed by Steve Jobs in 1980s.
- Numerous other modern operating system such as Linux, Unix, iOS, MacOS, Android, and many others are a few prominent examples of this era.

Application Software Generations

The development of modern software applications started from building automated codes for a particular function, mostly mathematical calculations, and deploying directly to the computers to get the desired results. Before that, the concept of software applications was not in the existing shape. Mechanical and electrical switching, signaling, and circuits were the main tools for providing certain instructions to the computer (then called 'calculators' interestingly!). The relationship of the advancements in the development of computer systems and the input mechanism are closely related to the evolution of modern application software development. Another big factor that catalyzed the software development was the emergence of modern computer programming languages developed by different researchers, businesses, corporations, and engineers. In other words, we can say that the classification of the generation of software application development is mostly based on the varying level of abstraction pertaining to computer languages, operating systems, and advancements in code development. The most fundamental component that plays very critical role in the definition of the generations or categories of the software eras is computer programming languages and the emergence of tools that help develop the software code for deploying them on the computer machines. Thus, the evolution of application software eras based on the level of abstraction can be loosely categorized into the following major generations [28].

First Generation

The first-generation software programs can be classified as the mechanical and electrical algorithms to instruct the computers/calculators to perform certain computations. This era starts from the creation of the first algorithm by Ada Lovelace in 1842 for the Analytical Engine. A range of new algorithms that mark time period before the development of machine code directly deployed to the machines can be marked as the first-generation of software applications.

Second Generation

The development and deployment of machine-readable code in different forms that require high-level of expertise, skill, and understanding of the codes is considered as the second generation of the computer software applications. This era pertains to the prior-era of the development of operating systems. The software application to run Manchester Baby in 1948 can also be counted as the second-generation software development [29].

Third Generation

The 3rd generation software applications may relate to the compiler programs and computer languages that relate the compiling of the specific tasks written in the computer programs into the machine languages that a computer understands. This era includes numerous software programs such as A-0 and related series developed by Grace Murray Hooper [30]. The other major software applications include FORTRAN, COBOL, LIST, and so on. Most of the software applications developed in this generation or era of software development were based on the procedural software development, which had certain limits in terms of growth, enhancement in computer capabilities, improved technologies in the modern electronics and many others. This era lasted until the end of 1970s and started to fade away when the new model emerged for software application development powered by the latest software platforms, languages, and tools to match the pace of emerging ICT technologies.

Fourth Generation

With the introduction of "VisiCalc" by Apple Inc., in 1977, it marked the advent of the modern software development that could cover the general-purpose utilities that can be easily used by the people with relatively lesser skills and expertise in computer programming. This opened up the new era for different applications for office, professionals, and businesses. The materialization of fourth generation software applications was powered by the high-level programming languages such as C, SQL, and others that pave the way for development of domain-specific applications to automate different processes without any large-scale intervention of human and other resources on computers. Introduction of a range of high-level and multipurpose languages revolutionized the software development. New applications were developed for standalone as well as connected environments. With the advent of modern Internet, the software applications took a new height in the field of information and communication technologies across the globe. The main examples of software applications that were launched in the beginning of this era include Microsoft Excel, Microsoft Word, AutoCAD, and many other professional and office applications [29, 30].

Fifth Generation

The growth of software development continued as the advancement in computer technologies and software development languages, platforms, and tools kept emerging in the marketplace. The interactivity, resource-reusability, code-reusability, responsiveness, backend & frontend scripting, real-time response, and many other additional features of modern software applications characterize the fifth-generation. The present-day web applications, mobile applications, domain-specific apps, knowledge-based automated processes, and customized/personalized applications development are the major characteristics of the 5th generation software development. The addition of artificial intelligence (AI), knowledge-based automated processes, real-time processing capabilities, cloud computing capabilities, distributed processing, and multi-location storage and other capabilities have become the fundamental components of modern software applications of this generation.

Programming Language Generations

Programming languages stand at the core of the entire revolution of modern information and communication technology worldwide. A programming language helps engineers create a range of software codes to instruct computers to perform numerous functions and activities that can bring great value to the humanity, society, and businesses. The development of modern operating systems for mobile phones as well as for other machines (not-limited to just computers such as switching devices, home appliances, automobiles, networking devices, and software-based infrastructure and many others) is fully dependent on the computer programming languages. The development of modern programming languages, tools, and platforms is also dependent on the computer programming languages!

Programming languages have evolved through the ages based on the development and research in different eras. Thus, the evolution of programming languages can be classified into multiple generations based on the power-level, capabilities, and features of the programming languages, as mentioned below [31, 32].

First Generation

The first-generation programming language is known for its characteristics of machine-level programming category. This group of programming languages did not use any compilers or assemblers for the code to deploy. Those languages would deal with directly machine-level programming through front-panels of the computers. These languages would deal instructions based on 1s and 0s, which required higher level of intellectuality and domain expertise.

Second Generation

The 2nd generation computer programming languages are also referred to as assembly languages. These languages would use longer correspondence between instructions created in language and the architecture's machine code instructions. The languages in the second-generation are also known as low-level programming languages. The main examples of the 2nd generation computer languages include RISC (Reduced Instruction Set Computing), CISC (Complex Instruction Set Computing), X86, ARM (Advanced RISC Machine), MIPS (Microprocessor without Interlocked Pipelined Stages), and others.

Third Generation

Third-generation computer programming languages are more sophisticated, capable, feature-rich, and computer-independent or portable. These languages offer many additional functions over the second-generation languages. The main characteristics of third-generation computer languages are mentioned below:

- Highly programmer-friendly and portable
- Enhanced support for expressing concepts and aggregate data types
- More abstract than the previous generations
- They support major traits such as object-oriented and structured programming
- Examples include FORTRAN, COBOL, C, Java, Pascal, Perl, PHP.

Fourth Generation

The most common characteristic of the fourth-generation computer programming language is the capability of domain-specific programming of applications. The programming languages dealing with different software development domains such as web development, database management, mathematical optimization, report generation, and others are referred to as the 4th generation programming languages. The examples of the fourth-generation languages include R, Oracle Report, PL/SQL, SQL, ABAP.

Fifth Generation

The 5th generation computer programming languages are characterized by the following features, traits, capabilities, and characteristics:

- Capable of solving problems based on the constraints to the computer program not on the basis of algorithm; thus, it has additional intelligence and capabilities
- Also referred to as declarative language and constraint-based languages
- Sometimes, known as logic programming languages

- Capable of using artificial intelligence (AI), machine learning (ML), and other advanced technologies into the code development automation such as artificial neural networks, domain-specific functional declarations, etc.
- Offer faster communication, greater efficiency, simpler models of communication, and many additional capabilities related to effectiveness, intelligence, and performance.
- Allow simple communication between computer machines and programs
- The examples of the 5th generation languages include OPS5, Mercury, Prolog.

Wireless/Cellular Technology Generations

Wireless technology is one of the most critical components in the modern field of information and communication technology (ICT), which has pivotal role in the advancements of the modern communication technologies. Wireless communication has a wide range of applications in our modern technological and business arena. A few of those domains include satellite communication, radio broadcast transmission, Wi-Fi communication, terrestrial and extra-terrestrial microwave communication, Bluetooth communication, NFC (Near-Field Communication) technology, wireless cellular mobile communication, and others. Different wireless technologies operate in different frequency spectrum and have different applications. The popularity of cellular wireless technology is so high in modern wireless communication because of its support for mobility and anywhere, anytime communication (if the infrastructure is in place). Due to the power, popularity, and capabilities, cellular wireless technology has evolved through different ages commonly referred to as the generations of cellular technologies. Let us learn about them here [33, 34].

Zero Generation

Zero generation wireless technology is pre-cellular wireless technology. The main example of zero generation wireless include walky-talky, wireless telephone extensions, and similar applications. There was no greater level of mobility between two cells in that case. The mobility would affect the performance and the Quality of Service (QoS).

First Generation

The first-generation of wireless cellular technology is characterized by the use of analog wireless signals. The analog cellular standards, mostly introduced in the late 1970s and early 1980s, would support analog encoding of audio signals over the wireless carrier at different frequency bands. The examples of the first-generation cellular technology include Advanced Mobile Phone Systems (AMPS), Total Access Communication Systems (TACS), and others. These standards used Frequency Division Multiple Access (FDMA) technology and would operate at 800 MHz with a voice carrier channel of 30 MHz each.

Second Generation

The second-generation cellular wireless technology is digital technology that operates in the same frequency band as the previous 1G technology did with additional digital technology for modulation. The main characteristics, features, capabilities, and examples of 2nd generation cellular wireless technology are listed below:

- Uses Time Division Multiple Access (TDMA) digital technology
- Supports multimedia messaging, text messaging, and low-speed Internet
- Operates at 900 and 1800 MHz frequency bands
- The main example of 2G technology is Global Systems for Mobile Communication (GSM), which was launched in 1991 in Finland.
- The advancements in 2G technology were made by adding GRPS (General Packet Radio Service) and EDGE (Enhanced Data rates for Global Evolution) networks for better Internet or data services.

Third Generation

Third-generation cellular technology provided greater speeds of Internet as compared to the previous generations. The main wireless standards for the third-generation wireless technology include Universal Mobile Telecommunication Systems (UMTS), IS95, cdmaOne, and CDMA2000. The additional standards for enhanced data transmission included 1xRTT, 1XEv-DO, and UMB (Ultra Mobile Broadband). Different versions of High-Speed Packet Access (HSPA) protocols were used for improved broadband service. The first 3G network was rolled out in 1998.

Fourth Generation

Long Term Evolution, precisely known as LTE, was the core technology that marks the start of the fourth-generation cellular wireless. The main features, characteristics, capabilities, and examples of the fourth-generation wireless include:

- The main standard for the 4th generation wireless is defined under 3rd Generation Partnership Project (3GPP) in two different versions or releases—8 and 9

- The introduction of Wi-Max technology is also referred to as 4G technology with both fixed and mobile capabilities of high-speed Internet
- The first 4G network was launched in Norway in 2009
- It operates at 700 MHz band and supports MIMO and OFDM technologies.

Fifth Generation

The 5th generation wireless network is fully based on Internet as the core network and offers more than 100 times better performance than the 4G network. It can offer broadband speed up to 10 Gbps. The other features and characteristics of this technology include:

- Offers higher capacity, lower latency, greater bandwidths.
- Uses millimeter waves, which are shorter than microwaves. Two frequencies are defined as FR1 (<6 GHz) and FR2 (millimeter waves).
- FR2 band may range between 24 GHz through 54 GHzh.
- The specifications of 5th Generation network are defined by the Next Generation Mobile Network Alliance (NGMN) Alliance.
- Three main characteristics specified under 5G technology include [35]:

    - Enhanced Mobile Broadband (eMBB)
    - Massive Machine Type Communication (mMTC)
    - Ultra-Reliable Low Latency Communication (URLLC).

- It uses the new radio interface referred to as the Fifth Generation New Radio, precisely known as 5G NR interface.
- This network is very effective for the Internet of Things (IoT), industrial, home and office automation and many other high-speed broadband connectivity.

WWW Generations

World Wide Web, precisely referred to as WWW or W3, is an interconnected ecosystem of electronic resources in the shape of webpages, which are accessible through active hyperlinks through the Internet. It is also known as "WEB". This is the backbone of the modern online resources that can be navigated with the help of hyperlinks on the webpages. In very simple words, a WWW environment is the platform for resource sharing, human interaction, collaboration, business coordination, machine and cognitive collaboration through modern Internet and related web technologies, and software-based resources. The resource on the Internet in the WWW ecosystem is recognized by a logical identity known as Uniform Resource Locator (URL). The World Wide Web environment has evolved through different eras as mentioned below [36, 37].

Web 1.0

Web 1.0 is the first-generation of WWW ecosystem, which is characterized by the following main features and capabilities:

- It is a read-only web environment in which only static webpages are used to provide visitors with only reading, listening, and viewing of the content posted by the owner of the content and webpages.
- No interactivity is supported in web 1.0 system.
- Proposed by Tim Berners-Lee, an English computer scientist.
- This ecosystem spans from 1996 through 2004.
- It is read-only, one-directional, and static content.
- Acts like a message board or buddy list (one-way communication).
- Offers higher control and content security to the owner.

Web 2.0

Web 2.0 is known for its capability to support interactivity among the users and content owners simultaneously like a community portal. The main features and characteristics of web 2.0 ecosystem are:

- It is the 2nd generation web environment proposed by Dale Dougherty in 2004.
- It is referred to as read–write web system or interactive system.
- Allows gathering and managing large number of people or crowds across the globe on a common interest platform to interact with each other by sharing, collaborating, cooperating, and communicating simultaneously on one single web platform.
- The users have less controls and more interactivity.
- The examples of Web 2.0 system include RSS (Really Simple Syndication) feeds, social networks, blogs, Wikis, Podcasts and many others collaborative platforms.

Web 3.0

Web 3.0 is an advanced generation of World Wide Web system that is envisioned for the higher level of interactivity among the users, machines, software applications, and other sensor devices integrated into a gigantic network of networks. It is also known as 3rd generation of web ecosystem for advanced Internet use. The main features, capabilities, and characteristics of World Wide Web 2.0 ecosystem are listed below:

- It is a semantic web system, which allows machines and software applications to increase their interactive interventions and reduce the tasks and decisions made by human manually.

- Support machine read-able content with greater level of interactivity by the hardware as well as software-defined apps and machines.
- It supports two main platforms—social computing ecosystem & semantic technologies.
- Web 3.0 has got grounds from 2016 and onwards.
- This web can also be known as executable web system with trillions of users both machine users and human users simultaneously.
- Supports latest technologies such as artificial intelligence, machine learning, 3D computer vision technologies and others.

Web 4.0

Web 4.0 is a futuristic web environment that has not yet got materialized at full potential. In this web ecosystem, extensive interaction, collaboration, and communication among machines, human, and applications will take place. This is the fourth-generation or the next generation World Wide Web environment, which will act like Ultra-Intelligent Electronic Agent on the Internet. It will also act as ubiquitous and symbiotic web agent. It will be read-write-concurrency web in real-time and real-world.

Evolution of Storage Technologies

The data storage technologies that we use currently play a vital role in the advancements of modern information and communication technologies. The evolution of data storage technologies started from mechanical storage through the latest storage technologies based on solid state devices and modern optical storages. Let us have an overview of the evolution of the major data storage technologies used in ICT [38, 39].

Initial Storage Technologies

As far as the initial data storage technologies are concerned, they date back to the eighteenth century when mechanical devices would be used for data storage, conveying information or instructions to the machines or other persons. For example, the first-time textile looms were controlled through punch card by Basile Bouchon in 1725. The punch card technology was also extensively used in 1837 in Charles Babbage's Analytic Engine calculator. The use of punch cards continued even till 1980s. The other initial storage technologies include:

Magnetic Drum Memory—Fast forwarding to 1932, the magnetic drum was first invented by Gustav Tauschek. One magnetic drum of 16-inch length could store about 40 tracks. Magnetic drum has long magnetic coated cylinders and read–write heads that would rotate at about 12,500 RPM. The first such drum stored about 48 KB data in.doc format.

Williams-Kilburn Tube—Williams-Kilburn tube is the primary storage technology based on vacuum tube technology. It could store 128 bytes. It was used as random-access-memory in the computing systems. This vacuum tube storage device was developed in 1947. The size of this device with cathode ray tubes was 16 and ½ inch long and 6 inches wide.

Magnetic Core Memory—The magnetic core memory device consisted of tiny magnetic rings, referred to as cores to store one bit of data on it. This magnetic core memory device was developed in 1951 that could store as much as 2 KB data. This storage became computing standard from 1955 through 1975.

Magnetic Tape-Based Technologies

Magnetic tape-based technologies got ground well in the beginning of the twentieth century. One of the tape-based storages was first developed in Germany in 1928. It was used in computing systems in 1951 when Eckert-Mauchly UNIVAC I was invented. This technology used motors to rewind the tape from one pulley to another one passing through a read–write head. It remained in use for a very long period till the recent years or even in some old technologies at present too. A few other tape-based technologies include the following:

UNISERVO Tape Drive—It was a metallic tape commercially used with UNIVAC system in different versions such as UNISEVO I, II, and III. It consisted of thin-plate of nickel and phosphor bronze. The size of this tape was ½ inch wide and 1200 feet long. It could store 128 bits per inch of the tape.

Whirlwind Core Memory—The Whirlwind core memory is a form core storage memory developed by the MIT Whirlwind Computers Inc. This used the same cores or rings of magnetic cores to store data on the device. It was used for many years in the computing.

DEC Tape—This is another form of data storage that used the magnetic tape technology. This storage device was created by Digital Equipment Corporation, precisely known as DEC. It was ¾ inch wide and formatted into multiple individual blocks, which were directly written or read by the heads. Each formatted block could store 128 K 12-bit PDP-8 words or 144 K 18-bit PDP-8 words. It was extensively used for a numerous DEC series of computers for a long period of time [40].

Magnetic Disk-Based Technologies

Magnetic disk-based storage devices use the magnetization process for writing and reading the data on the magnetically coated disks. The data on the magnetic disk is stored in the forms of sectors, tracks, and spots. It consists of a rotating magnetic platter and a mechanical arm to read/write data at different locations based on sectors, forms, and tracks [41].

Magnetic disks can be classified into two major categories based on the material used in the formation of disks such as:

- Magnetic disks made of flexible plastics
- Magnetic storage disks made of rigid glass or aluminium.

A large number of different types of storage devices were manufactured with a little variation in their respective technologies during the past few decades. A few most commonly used magnetic disk-based storage are mentioned below:

Card Random Access Memory (CRAM)—Card random access memory (CRAM) is a flexible plastic disk-based storage technology. It was extensively used for the mainframe computers. It first appeared in 1962 and was invented by National Cash Register (NCR) Corporation.

Disk Cartridge—Disk cartridges can be both rigid metals or optics. The disk cartridge uses a disk of magnetized material that is permanently housed within a protective cover made up of plastic cover. Disk cartridges can be either vertical loading or horizontal loading into the hubs where disk spindles reach the cartridge to rotate. The capacity of disk cartridges was 50 megabytes. It was first introduced by IBM Corporation in 1964 [42].

Disk Storage Drive—The disk storage devices use different types of storage disks in different forms and formats. Most of the types of disk storages include:

- Hard disk drive
- Floppy disk drive
- Zip disk drive.

Hard Disk Drive (HDD)—Hard Disk Drive, precisely known as HDD is a non-volatile data storage device that uses the principle of electromechanical data storage. Rigid platters are coated with the electromagnetic material. The platters are associated with the heads for reading, writing, and managing the data on the platters. HDD was first introduced by IBM in 1956.

Floppy Disk Drive—Floppy disk is small storage device for transferring data from one computer to the other one manually. This uses a thin disk of magnetic storage device protected with a plastic cover. Floppy diskettes were very popular from 1970 through 1990s before emails took over the file attachments between two computers or destinations. Again, when flash drive or USB pen drive arrived in the market, the need for floppy disk drives almost vanished.

Zip Disk Drive—It is a portable data storage device that was introduced by Iomega in the mid-1990s. It was introduced in different sizes such as 100 MB, 250 MB, and 750 MB. They continued to remain popular for medium sized data transfer until the latest storage devices such as USB drives were introduced in the marketplace.

Transformer Read Only Storage (TROS)—This is one of the earliest read-only-memory storage devices. It consisted of the printed wiring sheets to store the microcode used for running mainframe computers. It was introduced by IBM for different models of IBM Systems/360.

Semiconductor-Based Storage Technologies

Semiconductor-based storage technology uses different forms of semiconductor materials such as flip-flops, transistors, capacitors, resistors, polarization of ferro-magnetic material, gates, and others to store data in both volatile and non-volatile fashion. Thus, semiconductor-based storage technologies can be used for random-access-memory (RAM) as well as read-only-memory (ROM) [43]. A few major types of semiconductor-based storage technologies are mentioned below [44].

Static Random Access Memory (SRAM)—SRAM is a type of random-access-memory (RAM), which uses capacitors for storing the charges in such a way that they represent zero (0) and one (1). It does not require the additional refreshing to hold the charge as is required in the dynamic RAM. The most common application of SRAM is used in the cache memory in our computer devices.

Dynamic Random Access Memory (DRAM)—The capacitors used for handling charges for describing the digital value like zero (0) and one (1) need additional continuous refresher dynamically for maintaining the data. As the logical value is described with the charge value on capacitors, which tend to discharge; a dynamic refreshing of the data is required. Thus, this type of RAM is referred to as dynamic RAM. Most popular RAM technologies are:

- Synchronous Dynamic Random Access Memory (SDRAM)
- Double Data Rate (DDR) version 1, 2, 3, & 4.

Electrically Erasable Programmable Read Only Memory (EEPROM)—EEPROM is a type of ROM memory that is based on non-volatile memory principle and uses the semiconductor material for storing the data. It uses an array of floating-gate transistors to store data that is electrically erasable and programmable. It can be retrieved even after powering off the device. This is an advanced technology of EPROM and PROM technologies.

Flash Drive—Flash drive is a type of data storage technology based on the EEPROM technology, which uses the floating-gate transistors (MOSFETs) in the storage circuitry. The technology can be divided into two major categories: NAND gate-based and NOR gate-based technology. This technology was developed in 1980 by

Toshiba Inc. This storage device was launched in 1987 commercially [45]. It can be rewritten and erased for about 10–100 K times without any degradation of device quality. Flash drive has been very popular for almost over two decades now. The capacity, performance, and size of the flash drive storage are improving continuously.

Solid State Disk SSD—Solid State Srive, precisely referred to SSD, is a type of digital data storage device that uses electronic integrated circuits (ICs) for storing, accessing, and erasing the digital data. The data on SSDs is stored on semiconductor cells. This does not use any kinds of mechanical or electromagnetic materials like HDD and floppy disk would do. This technology is used for non-volatile data storage. It is used in both portable flash and hard drive storage devices because it is compatible with HDD in operations and interfacing for hybrid storage.

Optical-Based Storage Technologies

Optics-based storage technology uses low intensity beam of laser and polished optical disk. Low-power laser beam encodes the digital data on the optical disk through energization of tiny pits arranged in spiral track. The reading of the data is also based on the reflection of the light from those tiny pits to generate electric signals. A range of optic-based data storage media are used in modern computing such as [46, 47]:

- CD ROM—Compact Disc Read Only Memory
- DVD ROM—Digital Versatile Disc Read Only Memory
- DVD RAM—Digital Versatile Disc Random Access Memory
- HD DVD—High-Definition Digital Versatile Disc
- Blu Ray—High-Definition and Large Capacity Optical Discs
- CD-R/W—Compact Disc Re-Writable
- DVD-R/W—Digital Versatile Disc Re-Writable.

Advanced Storage Technologies

Initially, the storage was directly attached to the computer. But, with the passage of time, the networks emerged and the sharing of the data storage among the networked computers began. When the large and complex networks started sharing and creating huge volumes of data, the newer technologies started emerging in the ICT marketplace. Eventually, innovative storage media, techniques, and configurations emerged. The advancements in storage technologies and mediums have already been discussed in this chapter. Now, let us focus on the advancements in the configuration, sharing, storing, and accessing technologies [48, 49].

Direct Attached Storage (DAS)

The storage devices such as Hard Disk Drive (HDD), Solid State Drive (SSD), or optical storage devices connected directly with the computers internally or externally are referred to as directly attached storage systems. The directly attached systems are available for only one machine to which that storage is attached to access and store. It can be either any device such as optical, magnetic, or other but in most of the cases, these devices are HDDs and SSDs. The main interfaces through which DAS is connected include:

- Serial Advanced Technology Attachment (SATA)
- Small Computer System Interface (SCSI)
- Peripheral Component Internet Express (PCIe)
- Serial Attached SCSI (SAS).

Network Attached Storage (NAS)

A centralized node of data storage with independent IP address connected in a network for providing the shared data storage for different computers on a network is known as network-attached storage configuration (NAS). In this structure, different computers can access the network-attached-storage node like a server. This server resides on the local area network independently with adequate management and operations options related to data security, sharing, and other functionalities to provide the best possible solution. NAS devices can support a range of protocols; a few of those protocols are listed below:

- Common Internet File System (CIFS)
- Network File System (NFS)
- Server Message Block (SMB).

Storage Area Network (SAN)

Storage-area-network, precisely known as SAN, is a network of data storage machines with extensive management, operations, and security features. In this structure, many data storage servers are connected into one pool of data for faster access and efficient management. SAN networks can be formed with HDD, SSD, or both in a hybrid configuration. SAN networks use different technologies and protocols such as:

- Fiber Channel Configuration
- Internet SCSI (iSCSI—Internet Small Computer System Interface)
- Fiber Channel Over Ethernet (FCoE).

Futuristic Storage Technologies

With the incessant improvements in the storage technologies and other ICT technologies, new models and ideas surface continuously. The major data models such as NAS (Network-Attached Storage) and SAN (Storage Area Network) have their own respective limits and constraints, which force the businesses and researchers to develop new strategies that could overcome those limitations or downsides. To overcome the limitations of different data storage solutions, a few new concepts are taking the center point in the modern data storage systems as mentioned below:

Software Defined Storage (SDS)

Software-Defined Storage (SDS) is a comprehensive software layer that is used to separate the storage hardware equipment from the storage software that manages the data storage resources. It can handle different functions such as capacity provisioning, control management, data management, data protection, and other functions without any dependency on the underlying storage hardware vendors and configurations. The main features and characteristics of software-defined-storage systems are mentioned in the following list [50]:

- SDS Separate the underlying storage hardware from software
- Seamless integration with any storage hardware irrespective of vendors
- It Combines numerous (diverse) storage devices into a central storage pool.

Storage Virtualization

Storage virtualization is another very powerful model that is used in the modern data storage ecosystem, especially in the modern hybrid cloud environments. Storage virtualization is a process to accumulate multiple data storage devices or underlying storage hardware to create a large pool of data and make it available for separate machines or servers as one single storage.

Software Development Generations

The process of developing software code or programs, commonly referred to as applications that can be either system application or utility application, has started evolving after the 1950s–1960s. At first, the digital coding started in those decades for providing the instructions to the computers. There was no standard programming language or development methodologies defined at that time. With the course of time, the software development also progressed. The progress of software development era can be classified into three major generations such as [51]:

- Conventional Era—(1960–1970)
- Transition Era—(1980–1990)
- Modern Era—(2000–Present).

Conventional Era—(1960–1970)

This era is mostly dominated by highly skilled and expert computer programmers; that is why, it is also referred to as craftsmanship era. This is 1st generation software development period which is characterized by the following attributes:

- Powered by the custom tools, processes, and components developed by the organizations individually
- Most of those custom programs were developed in primitive languages
- Achieving desired objectives of software development projects was not up to the mark.

Transition Era—(1980–1990)

The second-generation of software development is marked as the software engineering era. This era is more sophisticated and advanced in terms of programming languages, tools, methodologies, and advancement in the ICT technologies. The main characteristics, features, and attributes of this generation of software development era include:

- Organizations started using more repeatable processes, off-the-shelf tools, and custom components in the range of over 70% or so
- The availability of commercial components was much lower than 30% such as operating systems, networking systems, database management, GUIs, and others
- The development of custom components was mostly done through high-level languages.

Modern Era—(2000–Present)

Modern era, alternatively referred to as the third-generation software development, is also termed as software production era. The development of software has become systematic and planned—this is indeed a full-fledged production era for software applications, products, and platforms. The main characteristics, capabilities, and attributes of modern era of software development include:

- Powered by managed and measured processes, automated environments, integrated ecosystem of the off-the-shelf tools
- More than 70% components, tools, and processes are available in the commercial marketplace
- Less than 30% components are required to be custom-built
- Very fast speed of software development powered by integrated automation, reusability, components (off-the-shelf), and methodologies.

Types of ICT Services

The developments of ICT technologies have occurred in different technical domains, which have already been elaborated in this chapter. The impact of those advancements on our day-to-day social and business lives is twofold. (1) transformation of the existing services, (2) emergence of new services. Both of these impacts resulted in the following major ICT services, which were either already-available and transformed into news shapes or emerged on the ICT arena.

Software Development

This is a new ICT service that did not exist before the advancements in the ICT technologies. At present, it plays the most pivotal role in the entire growth, existence, and future prospects of the entire landscape of information and communication technologies. Software development service is expanding drastically and new domains are emerging. A few domains include:

- Mobile application development
- Web application development
- Operating systems development
- Database development
- Custom application and process development
- Industrial automation and process management
- Open-source platform development
- And many more.

Computer Networking

Computer networking was among the primary services that emerged with the advent of modern ICT technologies, especially with the successful projects like ARPANET and others. At present, the computer networking has become a very fundamental part in globalization of the businesses by making the entire world connected through the most advancement networks providing high-level security, reliability, efficiency, and desirable business results. The domain of these services is continuously expanding with the continual advancements in ICT technologies such as software, wireless technologies, and cloud computing.

IT Infrastructure Management

After the advancements in software development and networking services, the management of IT infrastructure became more effective, remote, and global. This new domain of services emerged in ICT landscape and new advancements are also developing in this field to contribute to overall growth of ICT technologies.

Telecommunication

Telecommunication technologies already existed before the advent of modern digital information technologies. It started basically when Samuel Morse developed electric telegraph code in 1837. Presently, it has transformed into the fastest moving technology in the entire spectrum of technologies pertaining to ICT. It is transforming all businesses, processes, societies, cultures, and communities drastically.

Data Storage Service

Data storage service is one of the major commercial services that have originated from the advancements of information and communication technologies. It is a new domain that started from providing simple devices for data storage and continued expanding for many years. At present, the data storage services are powered by numerous modern business and technical models based on cloud, virtualization, web hosting, and many others. The storage services are continuously evolving to cater to the big data, which has just started to explode.

Storage Transfer Service (STS)

Storage Transfer Service (STS) is one of the latest services related to data storage management. This domain of service business deals with the migration of data from one type of cloud or provider of cloud to the other cloud or provider of data storage services seamlessly without any big downtimes, software codes, and transmission hiccups. The transfer of data is done in multi-cloud and private environments. The transfer of data from one location to another should comply with the basic security standards such as safety, reliability, integrity, privacy of the information. This domain of information technology-based service is continuously evolving by adopting new platforms, strategies, and software applications [52].

Database Management

Like many other process management services, database management is also a kind of process management service. It has become so important due to the creation of unprecedented volumes of data in the modern business processes and environments. The database management service started from just managing the valuable data of an organization that was critical for running the businesses smoothly but ended up in handling of heaps of new created data during the course of business operations. The present-day database management service has much wider scope as compared to a few decades back when database management as well as development services just started. The database development services can be merged into the software development domain of services. Database management services include:

- Designing of database architecture
- Fine-tuning of performance
- Security management
- Database system upgradation
- Database backup and recovery
- Patch management and monitoring.

Process Automation and Monitoring

Process automation service deals with the developing, designing, deploying, and operating the process automation system (PAS) in a range of industries. The process automation services provide manufacturing execution system (MES) and PAS services through comprehensive integration of a range of sensors, digital circuits, software applications, microprocessors, microcontrollers, and SCADA (Supervisory Control and Data Acquisition) architectures, remote terminal units, and other components [53].

The process of automation service existed long before the advent of modern digital technologies through different electric and mechanical feedback and control systems. Electromechanical-powered automation and standalone control system services evolved to the modern industrial automation services that encompass a range of industrial, office, home, and many other automation processes powered by the latest IoT technologies.

Data Analytics

Data analytics is the evolved form of "*The Study of Analysis*", which can be traced back in the eighteenth century. Initially, this service was not linked with any form of information and communication systems, but with the passage of time, this service has transformed into a completely new domain of service powered by the modern ICT technologies, platforms, tools, devices, and methodologies [54]. This got greater traction after around 1960s when the invention of modern computers happened and new systems started coming into existence, which could handle highly complex data to analyze and deduce the desired valuable information from a bulk of data. At present, the data analytics has become highly sophisticated service powered by the modern domain-specific software tools, artificial intelligence, neural networks, machine learning, and so on. Data analytics service has expanded to several types as mentioned in the list:

- Bigdata Analytics & Predictive Analytics
- Enterprise Decision Management
- Augmented Analytics & Cognitive Analytics
- Web Analytics & Retail Analytics
- And many others.

Cloud Computing Service

Cloud computing service is one of the latest services that emerged after the advent of modern ICT technologies such as mass storage, computer networking, high-speed transmission, and modern software platforms. Cloud computing services are based on three very fundamental business models which include:

- Infrastructure as a Service (IaaS)
- Platform as a Service (PaaS)
- Software as a Service (SaaS).

A wide range of modern IT services are provided through those basic models of cloud computing services. The examples of cloud services include virtual servers, distributed storage, commercial email services, online storage services, and so on. At present, the number of cloud computing services has grown exponentially; especially, under the SaaS model of cloud services.

Application Programming Interface (API) Service

Application programming interface (API) service is one of the fastest growing services that can create a big wave in the field of Software as a Service (SaaS) business model. Before diving into the API services, let us explore what exactly API is. An API is a software program that provides the details of how to interact with a system to access its resources. Different systems allow both custom and third-party APIs to access their resources. For instance, you can use any supported third-party authenticator for verifying your credentials on a particular service or system [55, 56].

A SaaS-based service that allows a user to build custom APIs as well as connect an application or platform through third-party APIs on a common platform is known as API services. Such software-based platforms allow customers to create, test, deploy, and market their respective customized APIs as well as use range of third-party APIs available on the platform to connect to certain backend software-based cloud applications or system resources. These APIs directly deal with the backend microservices that are containerized independent application components.

Cybersecurity Service

Cybersecurity (or, Cyber Security) field emerged from the advancements of ICT technologies and overshadowed the importance of all modern ICT businesses by making all those businesses dependent on this one. With the increased threat profile and risk associated with security breach, the size of cybersecurity service industry is expanding exponentially across the globe.

A comprehensive service of predicting possible threats, deploying preventive measures, monitoring threats continuously, and building after-attack security strategies is referred to as cybersecurity service. Cybersecurity service did not exist till 1986, when the first computer virus named as brain was detected. This opened up the arena of a new business, which is referred to as cybersecurity today. Within a short period of time, the size, shape, and landscape of cybersecurity has changed drastically. At present, cybersecurity can be divided into five further categories [57]:

- Critical Infrastructure Security
- Cloud Security
- Application Security
- Network Security
- Internet of Things (IoT) Security.

Malicious users, state and non-state actors, and professional hackers in different fields are becoming more and more skilled and using sophisticated approaches to compromise the security of online ICT resources. This creates greater concerns as well as great demand for advancement in the technologies related to cybersecurity.

Digital Entertainment Services

Digital entertainment is a wider canvas of different services. A few of those services evolved from the traditional entertainment services and a few of them emerged from the advent and advancement of the modern information and communication technologies.

The traditional entertainment services that evolved and transformed into an enhanced mode of services in the entertainment industry include the following [58]:

- Film and television services
- Video gamming.

These services already existed before modern revolution in ICT technologies. The emergence of modern technologies changed the shape of those entertainment services drastically. For instance, animation, computer-vision, interactive illustration, and other technologies improved the quality and user experience of film and television services significantly. Meanwhile, the use of modern software and telecommunication technologies changed the landscape of video gaming heavily. Newer technologies made them more interactive, responsive, and powered by the modern wearables and other gaming equipment such as Xbox and others.

The other forms of entertainment services that originated from the advancement of modern ICT technologies include the following:

- Live streaming
- Computer games
- Mobile games
- Online gambling
- Social media.

All these forms of entertainment services are fully dependent on the modern information and communication technologies as far as the development, enhancement, access, and interactivity are concerned.

Content Delivery Network (CDN)

The content delivery network is not a service but rather it is a network of distributed servers that provides the services of faster delivery of content to the nearest users. This technique improves the performance and user experience of the Internet users in the most competitive world of modern businesses based on the information and communication technologies.

Almost all major Internet and cloud service provider companies worldwide are using CDN networks to improve the performance, quality and effectiveness of the services. The emergence of cut-throat competition in the global market, enhance quality standards, and ever-demanding users have increased the demand for the CDN networks used in all cloud services for better impact on the globalization of the services.

A Peep into Next Generation Technologies

The next generation technologies are those techniques, platforms, algorithms, or processes that would transform the existing services and businesses and create new landscape of opportunities, advancements, and growth. The next generation technologies help the global community to get connected to a faster, reliable, and round the clock ecosystem of information and communication systems [59].

According to the renowned consultant company McKinsey Inc, the impact of the next generation technologies will be unprecedented on all domains of businesses and society, especially the trade and value chains [60]. The global value chains are going through a drastic transition by undergoing numerous structural shifts such as manufacturing is becoming less trade-intensive, services are becoming more pivotal, and global value chains are becoming more knowledge-intensive. Those structural shifts are emerging due to the increasing impact of modern information and communication technologies impacting all domains of business, society, and global structure.

There are many companies, countries, consortiums, and organizations that are focusing on the development of new standards for the upcoming technologies so that the future demands of business, society, government, and global organizations can be met easily. They are conducting extensive research and development (R&D) to achieve those objectives. The main examples of the next generation technologies that are expected to leave extraordinary impact on our modern lives as well as business processes will be covered in this book in the following chapters which include:

- Artificial Intelligence (AI)
- Machine Learning (ML)
- Blockchain Technology
- 5th Generation Wireless
- Internet of Things (IoT)
- Distributed Cloud Computing
- Quantum Computing
- Tactile Virtual Reality (TVR)
- 3D Printing
- Digital Twin Technology
- Mind Uploading Technology
- Autonomous Robotics
- Artificial Neuron Technology
- And others.

Sample Questions and Answers

Q1. What is a firmware?

A1. A firmware is a kind of software code that is developed for a small device-specific hardware to provide low-level control over the operations of that particular electronic device.

Q2. What are the categories of Operating Systems?

A2. Operating systems can be classified in a range of categories like:

- Single- and multi-tasking operating systems
- Distributed operating systems
- Single- and multi-user operating systems
- Real-time operating systems
- Embedded operating systems
- Templated operating systems
- Library operating systems.

Q3. What is fourth-generation computer programming language? Give some examples.

A3. The programming languages dealing with different software development domains such as web development, database management, mathematical optimization, report generation, and others are referred to as the 4th generation programming languages. The most common characteristic of the fourth-generation computer programming language is the capability of domain-specific programming of applications. The examples of the fourth-generation languages include R, Oracle Report, PL/SQL, SQL, ABAP.

Q4. State three (3) notable features of Web 1.0.

A4. Web 1.0 is the first-generation of WWW ecosystem. Three notable features of Web 1.0 are:

- It is a read-only web environment in which only static webpages are used to provide visitors with only reading, listening, and viewing of the content posted by the owner of the content and webpages.
- No interactivity is supported in web 1.0 system.
- Acts like a message board or buddy list (one-way communication).

Q5. What is Storage Transfer Service (STS)?

A5. Storage Transfer Service (STS) is one of the latest services related to data storage management. This domain of service business deals with the migration of data from one type of cloud or provider of cloud to the other cloud or provider of data storage services seamlessly without any big downtimes, software codes, and transmission hiccups. The transfer of data is done in multi-cloud and private environments. The transfer of data from one location to another should comply with the basic security standards such as safety, reliability, integrity, privacy of the information. This domain of information technology-based service is continuously evolving by adopting new platforms, strategies, and software applications.

Test Questions

1. What are the implications of next-generation technologies?
2. What are the five categories of cybersecurity?
3. What is an Application Programming Interface (API) service?
4. What is cloud computing?
5. What do you mean by 'data analytics'?
6. What is the need for database management?
7. What is computer networking?
8. How do you define storage virtualization?
9. What is the working principle of IT protocols?
10. What are the differences between software and hardware?
11. What are the classifications of programming languages?

Chapter 2
Artificial Intelligence Technology

Introduction

Artificial Intelligence, precisely referred to as AI, is one of the most powerful, prospective, and attractive domains of technologies in all modern businesses, processes, as well as in social and governmental activities across the world. AI is playing a very critical role in the development of modern software applications and platforms that enable them to incorporate human capabilities such as learning, deciding, and acting in accordance with the given conditions or environments. The most common areas that the present-day AI research and development (R&D) focuses include perception, knowledge, reasoning, planning, and communication. These domains have greater impact on the advancements of modern information and communication technologies. There are numerous applications of AI technologies, especially the process automation in almost all industries and domains of businesses that make it the most influential one in today's world [60].

What Is Artificial Intelligence (AI)?

Artificial intelligence is a technology that builds and enables machines or non-human entities to perform tasks that are the core characteristics of human beings. AI is the combination of mathematics and computer science to build a software or hardware/software-based machine such as robots to perform the most important cognitive activities that a human performs. As the name indicates, the *intelligence* is based on the artificial technologies not the natural ones. In fact, there is some debate in the research community whether it could be called '*synthetic intelligence*' instead of artificial intelligence. In reality, it is nearly impossible to reach the human level intelligence which requires years of information collection and connecting the dots between meaningful information based on life experiences. Also, human beings often act based on natural tendencies or intuitions, which a machine cannot do in practice. A machine can run based on programming codes or logic that would be controlled by a set of instructions. Even if there is a learning mechanism, the entire process is still controlled by the machine's set of instructions introduced to the machine! The most common cognitive functions associated with the artificial intelligence include the following [61, 62]:

- Learning
- Problem-solving
- Reasoning
- Decision-making
- Speaking language.

What Is Neural Network?

Neural network is a network of neurons connected with each other. In artificial intelligence, it is referred to as Artificial Neural Network (ANN) or Simulated Neural Network (SNN). This network mimics the human neuron system that sends, receives, and processes the intelligence signals of the body and responds to the input accordingly. The schematic diagram to depict the neural nodes, layers, and connections are shown in Fig. 2.1.

Fig. 2.1 Neural nodes
connected in layered fashion
(Pixabay)

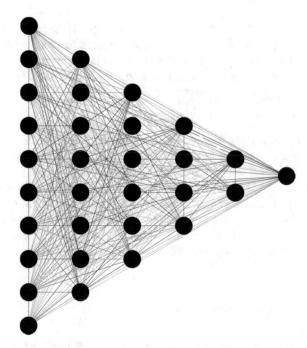

The most common characteristics, features, and structure of an artificial neuron network can be summarized as [63]:

- It is a network of nodes, connecting media, and different layers
- The names of main layers of nodes include:
  - Input layer
  - Output layer
  - Hidden layers.
- Node is equivalent to neuron in human brain analogy
- Hidden layers can vary; i.e., can contain multiple layers
- Each node has a threshold value to transmit signal above that value
- It sits at the core of the machine learning in the artificial intelligence
- Neural networks are algorithms that take training datasets for learning purpose.

Major Approaches Used in Artificial Intelligence Research

Artificial intelligence is hugely dependent on the capabilities, power, and the methodologies used for making the machines capable of learning about the real-world scenarios in such a way that they can make a decision in response to activities happening around the concerned surroundings of that particular machine or artificial intelligence application. In artificial intelligence field, learning/understanding, reasoning, and decision making are done on the basis of certain approaches that are adopted by the researchers in the modern machine learning. Those approaches are also evolving with the improvements in the technologies; but the basic approaches used for machine learning research are two as listed below [64]:

- Symbolic Approach
- Connectionist Approach.

Although, they are two different approaches with significant differences in their respective methodologies, algorithms, and techniques; there are certain conformabilities that make them so compatible and complimentary to each for producing more robust and desirable results in the machine learning research in many fields. In many cases, both of the approaches are integrated in a hybrid approach ecosystem for carrying out the most sophisticated and advanced engineering and cognition modeling of an artificial intelligence system. Let us define these two approaches separately to figure out the major differences, features, and procedures used in them.

Symbolic Approach

Symbolic approach is also referred to as Symbolic Artificial Intelligence, precisely SAI. The symbolic approach is based on the notion that the whole world can be taught to the machines through the input of symbols used to denote an entity in the world. The symbols are the fundamental units or building blocks of the entire world. The operations between the symbols are considered for making the machine algorithms understand the response to the situation when a query is put up by a user or automated system connected to the artificial intelligence-based machine.

The operation on the symbols is also referred to as the rules of operations. A large set of rules that govern the cognition through a large set of symbols is called as the expert system. An expert system is a set of commands or instructions based on *if–then*-based instructions. The expert system works on the knowledge base developed by the human experts. This knowledge base contains the information about the environment or world in the form of symbols. The inferring/inference engine or computer chooses the rules to apply to the particular symbols to generate intelligence. Thus, we can say that the symbolic AI modeling can be referred to as the mind modeling with presentation. This approach is very well suited for the fields such as robotics, automobile, telemetry, aviation, and others where the input data or the real-world information is based on the symbols or presentations. Converting those symbols into mathematical expressions/description or software code manually is not feasible as compared to the direct input of the knowledge through symbols to build a knowledge base for making the artificial intelligence algorithms understand/learn about the system through presentation of items and environment of real-world.

Connectionist Approach

Connectionist approach of artificial intelligence research is also known as Numeric Artificial Neural Networks, precisely referred to as NANN. According to this approach, the artificial intelligence can be developed through a large network of processors that work like neurons in a human mind to process the signals coming from multiple input sources. Each neuron is an independent node that is connected with all other neurons or nodes in the network and connected in the mesh form initially. Thus, the core definition of this approach to artificial intelligence research is that the decision-making in artificial intelligence can be achieved through a network of artificial nodes, i.e., interconnected system of small nodes.

In this approach, the value of activity (activism) of each node is defined based on the frequency of the signals processing. If a particular node is receiving multiple signals and processing, it is referred to as increasingly activated; otherwise, it is deemed to be inactivated. The size of neural network connection is determined on the basis of level of activation of the connected nodes. The complete structure of neural network is known as neural network. In connectionist approach of AI, the neural network is subjected to more information for learning. This network of convolution and transformation can learn the most complicated environments, especially in-depth models of data regeneration distribution, which lead to effective decision-making capabilities like classification, regression, and others, when trained with the desired data.

The most important characteristics and properties of connectionist (numeric) artificial intelligence approach include [65]:

• A set of neural or processing units of small sizes
• States of activation—activated or deactivated
• Predefined input functions
• Weight metrics
• Information learning rule
• Transfer function
• Model environment.

Objectives of Artificial Intelligence

The concept of artificial intelligence started getting traction among the scientists when they started benefitting from the improved performance, power, efficiency, and capabilities of machines that could accomplish tasks faster and more effectively. Those machines were more productive and faster than humans but lacked in the intelligence. If they could think and decide about the response to a given condition, the world would change drastically. A large number of business processes, activities, tasks, and even numerous creative things could be automated with the help of machines if they could learn, understand, and make decisions properly. To achieve those desires, the emergence of the concept of artificial intelligence occurred. The most common objectives that humans wanted to achieve through artificial intelligence included [66]:

- Making machines learn
- Making decisions powered by machines
- Solving complex problems faster
- Execution of multiple tasking
- Getting peep into the future
- Exploring areas beyond human-thinking.

All of the above-mentioned objectives can be achieved (to some extent) by the following processes implemented on the machines with the help of artificial intelligence:

Reasoning

Reasoning is one of the most critical objectives to achieve with the help of artificial intelligence. It is a logical process of developing an approach towards a thought, drawing conclusions, and making predictions. The reasoning plays the role of core component in making a decision. There are a few major types of reasoning used in the artificial intelligence:

- Deductive reasoning
- Abductive reasoning
- Inductive reasoning
- Commonsense reasoning
- Monotonic reasoning.

Problem Solving

Problem-solving in artificial intelligence refers to the techniques to solve a complex problem to achieve the most desirable results. This technique mimics the problem-solving process conducted by a human being. In artificial intelligence, a range of techniques are used for solving a particular problem such as heuristics programming, root cause analysis, and efficient formulation of goals and problems. The most fundamental steps used in AI problem-solving include [67]:

- **Goal formulation**—It uses finite steps to design goals to achieve, through certain actions. AI agents are mostly used for formulating the goals.
- **Problem formulation**—This step helps design the suitable actions to achieve the desired goals/objectives.

The most common factors that are used for formulating the problem in AI are mentioned in the following list:

- **Initial state**—The starting state of the problem from where the AI agent starts to achieve the desired goals.
- **Action**—This component performs all possible actions after initial state. It works with function with a specific class.
- **Transition**—The integration of action done in the previous stage and forwards to the next stage.
- **Goal test**—This factor verifies whether the specified goal is achieved or not.
- **Path costing**—This defines the numeric cost of the goals to achieve through entire problem-solving process.

Natural Language Processing

Natural Language Processing, precisely referred to as NLP, is a mechanism/process of machine learning or artificial intelligence. This process helps machine understand the spoken and written languages like a human does. This process consists of the following components [68]:

- **Computational linguistics**—It is used to model human language based on rules.
- **Statistical and deep learning**—These components enable machines to process human language in such a way that a comprehensive understanding of text, voice, and sentiments is accomplished properly.

The above-mentioned components are accomplished by performing a range of tasks in the entire process of natural language processing. A few of the major NLP tasks are mentioned in the following list, such as:

- Speech recognition task
- Natural language generation
- Pasts of speech tagging
- Word sense disambiguation
- Sentiment analysis
- Named entity recognition
- Co-reference resolution.

Learning

Learning is one of the most basic goals of AI. It is a process, which enables the machines to understand about the environment, which includes a range of information such as voice, objects, motion, blow, winds, sun, heat, temperature, and many other things. This process is similar to the capability of a human being to learn naturally about the environments that he/she comes in contact with. The machine learning can be categorized in terms of feedback mechanism in the classes as mentioned below [69]:

- **Supervised learning**—It is based on external feedback of environment in the form of input data mapping over the output.
- **Unsupervised learning**—Self-understanding by machines from the input data without any mapping or external feedback.
- **Semi-supervised learning**—This is a process like a middle-ground between supervised and unsupervised learning.
- **Reinforcement learning**—A method of learning based on rewarding if the desired behavior is achieved and punishing if undesirable behavior is achieved.

Planning

The most effective planning of the resources, activities, processes, and even of future times is another major objective of artificial intelligence. In the conceptual context of artificial intelligence, planning is a process of describing domains, goals, and action specification to achieve the desired objectives of an AI project. Artificial intelligence uses two levels of planning as listed below:

- Forward State Space Planning (FSSP)
- Backward State Space Planning (BSSP).

Both of the above-mentioned levels have their respective advantages and downsides for a particular planning project. A suitable level should be chosen in line with the description of domains, goals, and action specifications to achieve the goal stack planning.

Knowledge Representation

Knowledge representation is a process of artificial intelligence technology. It is used for presenting the information of real-world in such forms that machines can easily understand and learn from it and also perform response tasks automatically. There is range of information in the real-world scenario, which is to be translated in such a form that machines powered by AI technologies can understand easily.

Artificial intelligence uses different types of knowledge representation models. A few of them are mentioned below [70]:

- Logical knowledge representation
- Frame representation model
- Production rules
- Semantic network.

Motion and Manipulation

Detecting and controlling of motion and manipulating motion-based tasks through machines are some other major objectives of artificial intelligence. The operations and control of robotic arms with the help of AI software is known as motion and manipulation of motion.

Artificial General Intelligence

Artificial general intelligence, precisely AGI, refers to the understanding and suitable response of machines through intelligent agents to perform tasks related to general or common intelligence. This is one of the most desirable objectives of artificial intelligence and machine leaning of the modern world.

Social Intelligence

As far as the use of artificial intelligence is concerned in the modern world, the objectives of artificial intelligence have expanded significantly. Those objectives are overlapping many processes, tasks, activities, and hidden intelligence in different domains, industries, items, and entities. Social intelligence is one of such objectives associated with the artificial intelligence, where an intelligent agent can perform the same activities like a human being can. It can be expressed in this way:

> The capacity of machines to understand the social interactions and behaviors of other social elements and respond to them appropriately in terms of context, emotions, and actions is referred to as social intelligence.

Business Intelligence

Scanning different behaviors of markets, end-users, processes, and trends to dig out the most useful information that can help enhance business bottom lines is known as business intelligence. Due to increased volumes of data that describe the actions, events, trends, and other related-processes in the modern era of information technology, the use of artificial intelligence is very critical because managing such a huge data that bears valuable intelligence for enhancing business manually is usually very difficult.

Thus, another modern objective of artificial intelligence is to analyze big-data through different AI-powered components such as analytics tools, big data management platforms, robotics, and others. Those components can collect, store, access, analyze, process, and skim through the most beneficial information that can be very useful for businesses.

Machine Perception

Building machine perception refers to developing the capabilities in machines to sense, see, listen, taste, touch, and feel through the processing of different types of signals originated through different sensors. A range of metrics are used for training the machines to build perceptions regarding the above-mentioned senses. By developing such capabilities in machines, numerous applications can be built to automate a wide range of processes, activities, and functions easily.

There are a few very popular fields of machine perception in artificial intelligence that are mentioned below [71]:

- Machine Hearing
- Machine Olfaction
- Machine Vision
- Machine Touch.

An Overview of the History of AI

Myths play very vital role in scientific research and development (R&D). The same case is true for the development of artificial intelligence in the modern world. The myth of mechanical-men was very well known in ancient Egyptian and Greek cultures. The concept of mechanical men resembles to the artificial intelligence of modern era. Those myths started converting into different scientific fictions, plays, and futuristic ideas during the first half of the twentieth century and afterwards. The practical emergence of the artificial intelligence started in the middle of the twentieth century when the evolution of neurons and Turing Machines came to surface. The modern history of artificial intelligence can be deemed to have started from that era.

An overview of the modern artificial intelligence era timeline is given in the following list [72, 73]:

- 1943—Artificial neuron evolution
- 1950—Development of Turing Machine
- 1956—Official birth of AI domain via Dartmouth Conference
- 1966—Development of ELIZA, the first chatbot
- 1972—First intelligence robot WABOT-I was invented
- 1972–80—Downslide of artificial intelligence referred to as AI Winter-I.
- 1980—Development of Expert System
- 1987–93—Another downslide of AI known as AI Winter-II

- 1997—Invention of IBM Deep Blue, which beat the world champion in chess game first time.
- 2002—Invention of Roomba, the first for the home AI applications
- 2011—IBM Watson wins the quiz competition show first-time
- 2012—Development of Google Now
- 2014—Invention of Eugene Goostman chatbot, which passed the Turing test first time.
- 2015—Development of Amazon Echo
- 2020—Embodied Moxie, a robot companion for children.

This is very important to note that the concept and practical approach to artificial intelligence started to get mature during 1943 through 1950 when the first-ever AI-powered machine was developed. The concept of artificial neuron and Hebbian Learning emerged during this period, which lead to starting of the AI-Era.

The second era, which is considered as the birth-era of artificial intelligence is 1950–56. During this period, the Logic Theorist program was kick-started and the term "*Artificial Intelligence*" was also adopted. Two periods remained very grim for the advancement of artificial intelligence. Those two eras are referred to as Winter-I and Winter-II. Those eras spanned between 1974–1980 and 1987–1993.

Another very booming era of artificial intelligence started during 1980–1987. Earlier to this era, an enthusiastic era spanned over 18 years from 1956 to 1974. During the second boom-era of AI, the development of "*Expert System*" and holding of AI conference by American Association of Artificial Intelligence were two major events.

Intelligence agent era is considered between 1993 and 2011. During this period, numerous inventions took place such as IBM Deep Blue, Roomba vacuum cleaner, and integration of numerous intelligence agents into the services like Facebook, Twitter, Google, Netflix, and others. This era can also be marked as the new era of AI-powered software applications.

The modern era, which is much faster and advanced than previous period starts from 2011 and continues onwards. This era has seen numerous advanced applications and inventions related to artificial intelligence such as Google Now, IBM Watson, Eugene Goostman, Project Debater, and many others that are revolutionizing all processes, activities, applications, functions, and social behaviors of businesses, communities, and governments.

Main Areas of AI Application

A few decades back, artificial intelligence was mostly focused on robotics and industrial process automation, but with the advent of modern tools and platforms of software programming, the area of application of artificial intelligence has expanded exponentially. Before looking into the main areas of applications of artificial intelligence, let us have a look at a few research figures.

According to the forecast of Statista (at the time of writing this book), the artificial intelligence software revenue will cross $126 billion by 2025 [74]. Meanwhile, AI-Business research predicts that more than 95% of the customer interactions across the globe in all formats and forms will be powered by the artificial intelligence platforms or software applications [75]. These figures clearly indicate the exponentially-growing impact of artificial intelligence in a range of areas of applications. The most important areas where a range of technical applications of artificial intelligence are incorporated include [76]:

- e-Commerce and m-Commerce
- Education and Training
- Lifestyle and Equipment Management
- Navigation and Robotics
- Healthcare and Medicines
- Gaming and Social-Media
- Agriculture and Automobiles
- Finance and Banking
- Defense and Security.

Different types of algorithms, AI-technologies, models, processes, and platforms are applied in different areas in the above-mentioned list. A few very important technical applications, algorithms, technologies, and processes used in the above domains are explained below.

Natural Language Processing

Natural Language Processing, precisely referred to as NLP, is one of the most important domains of machine learning and artificial intelligence. It develops the capabilities in intelligence machines to communicate with the human through natural language that humans use for communication among themselves. This is a comprehensive area that uses a wide range of algorithms for different steps of processing the structure of naturally-spoken words, sentences, and paragraphs. Natural language processing intelligence systems can accept both speech and written text in different formats as an input and provide the desired output in those forms. Thus, an intelligent system powered by the NLP can communicate with users in text and speech irrespective of the form of inputs that the users are choosing. Intelligence systems can also translate text into multiple languages as well as transcribe speech into text and vice versa.

There are two major components or factors, which form the entire process of natural language processing technology such as [77]:

- Natural Language Understanding (NLU)
- Natural Language Generation (NLG).

The most common tasks performed by the natural language understanding (NLU) unit are more complex and require numerous algorithms and software processes as compared to the natural language generation (NLG) unit. It has to deal with a range of difficulties for understanding the natural language because it consists of numerous ambiguities, which could make the intelligent agent more confused in understanding the proper meaning, word differentiations, sentence meaning, formatting, tones, and so on. Those ambiguities are classified into three major categories based on the level of ambiguity in the sentences, phrases, or words:

- **Lexical ambiguity**—This is a type of error in natural language. It is also known as semantic ambiguity. This relates to the confusion regarding the meaning of a word. In some cases, one word has different meanings and in the other conditions, one word has numerous meanings and grammatical forms.
- **Syntax ambiguity**—This is a type of vagueness in natural language in which one sentence can provide two different meanings.
- **Referential ambiguity**—This is a type of vagueness in natural languages in which a word/phrase (in most cases pronouns) refers to two or more things or properties. This happens in different contexts of the sentences.

Thus, the NLU has to deal with different levels of vagueness used in natural language to perform numerous functions. The most common functions of NLU include:

- Mapping natural language-based input—either in text or speech form—into useful representations for building understanding of the input language.
- Analyzing numerous aspects of natural language for accurate understanding.

On the other hand, the main objective of NLG unit is to generate natural language-based meaningful sentences, words, and phrases, from some internal presentations. The most common functions performed by the NLG unit include:

- **Sentence planning**—Selection of accurate words to form meaningful phrases and setting tones of the sentences.
- **Text planning**—Retrieval of relevant content from knowledge-bases.
- **Text realization**—Sentence plan mapping into sentence structure.

All of the above-mentioned functions dedicated to each component of NLP have to go through five (5) fundamental steps. Those steps deal with different analysis, comparative, and integrative processes to understand and generate the suitable presentations as well as natural language in the form of written text and speech. They are mentioned below [77, 78]:

- **Lexical analysis**—Lexical analysis is also referred to as morphological analysis, which deals with the meaning, fractions, and grammar of a word. Breaking a word into its components such as prefix, suffix, and root and then, assigning the grammar tag such as verb, adjective, or noun, etc. are the most common functions of this analysis. Mostly, the lexical analysis uses the NLP dictionary for analysis of a word for reference.

- **Syntactic analysis**—This step analyzes the meaning of the sentence and assigns the parts of speech tags in terms of correct sentence. It makes sure that the sentence structure is correct based on the predefined rules and grammar. Any sentence that does not fit within predefined rules is discarded or marked as incorrect. Thus, we can say that it is a method of parsing the sentence for correct sentence structure (called syntactic/syntax analysis).
- **Semantic analysis**—This step analyzes the sentences for their accurate meanings and also groups the names of person or any other category. The combination of naming words to form a phrase is one function of this analysis; and analyzing the meaning of the sentence is the other function of this step.
- **Discourse integration**—In this step, the connection of the previous sentence is established with the next sentence to create an effective referencing and building a correct context. It is very important to note that the previous sentence has an effect on the coming sentence in a natural language. For instance, two sentences like, *"Mr. X is a great orator. He has command over many fields of knowledge."* In this sentence, *"He"* connects with *"Mr. X"* in the first sentence to build a smooth flow for an integrated discourse.
- **Pragmatic analysis**—As the name implies, pragmatic analysis generates the real-world meaning of the sentence analyzed in different ways as mentioned in the above steps. It finalizes whether, the sentence is a command, a request, a description, or any other natural associated with the effective expression of ideas in natural language.

The above-mentioned 5 types of analysis are the stages of natural language processing for a machine to learn and respond from the given input. The most crucial stage of this entire processing is syntactic analysis, which lays the foundation for learning of the components of the sentence based on the right structure and grammatical attributes. It is done through a wide range of algorithms developed by different organizations and professionals. A few very important algorithms commonly used for syntactic analysis include:

- Context-Free Grammar (CFG)
- Top-Down Parser.

Computer Vision

Computer vision, precisely referred to as CV, is an artificial intelligence area, which enables the machines to see, observe, and understand the information through an image or a video. This is a very complex procedure of data science research in the field of artificial intelligence and machine learning that can scan knowledge from visual frames and act like a human does by looking at the environment around and deciding for an action.

Computer-vision is similar to human-vision, which is capable of understanding the environments through videos and images, and can analyze, understand, and decide about the distance of an object, motion of that object, direction of motion, faults and features of an image and many other things. The computer-vision process is used to train the computers through a range of images and videos to understand and decide about all aspects of an image that a human-vision can do. The main functions performed by a machine through computer-vision technique include [79]:

- **Object recognition**—This task addresses three main issues associated with object differentiation such as object detection, object classification, and recognition of a particular object such as face-recognition and others. The most common tasks performed in object-recognition category include:

  - Pose estimation
  - Content-based image retrieval
  - Optical character recognition (OCR)
  - Bar-code reading
  - Shape recognition technology
  - Facial recognition.

- **Motion detection and analysis**—The motion detection and analysis are a procedure in which an image sequence is construed to generate the direction and speed of motion of an object. The major sub-tasks of this category include:

  - **Motion tracking**—Following the object through different tracking techniques.
  - **Egomotion**—detecting 3D rigid motion such as translation and rotation.
  - **Optical flow**—This detects the motion of different points within an image and also the apparent motion of the image as a whole.

- **Scene reconstruction**—Using the technique of point cloud and other 3D models to construct a 3D scene from multiple point cloud and grids taken from multiple angles is known as reconstruction of 3D image scene.
- **Image restoration**—This approach is used to remove different types of noises from an image with the help of different techniques.

Computer-vision technology is highly influenced by numerous complex algorithms coded through software applications, which are able to perform the following tasks:

- Image acquisition
- Image pre-processing
- Feature extraction
- Image detection and segmentation
- High-level image processing
- Decision making.

Expert Systems

We talked about expert systems briefly before. Expert system is basically another very important area in the artificial intelligence field. An expert system is a machine that is able to take decisions like a human does. This system may contain one or more computer machines and powerful software applications that make it efficient combination to make decisions like human in the given conditions (apparently, like a human being). The application of expert systems includes the area where complex problems related to certain business and scientific processes are supposed to be solved with the help of intelligent-system that is able to make decisions in certain conditions. Traditionally, these machines use *if–then* rules for reasoning purpose to make a decision.

The main characteristics, features, and capabilities of an expert system can be summarized with the following points [80, 81]:

- Computer system that can make decisions like a human being (seemingly, though not exactly likea human being).
- Automatically solves the most complex problems that need decision-making power frequently.
- It depends on two main components—Inference Engine and Knowledge-base.
- A knowledge-base consists of the facts, figures, data, procedures, and related information while the inference engine is a decision-making entity that evaluates, assesses, analyzes, and decides about the required actions.
- The most common tasks associated with the expert system include monitoring, classification, diagnosis, analysis, scheduling, designing, feedback and controlling functions, interpretation and prediction capabilities, and specialized planning.
- *If–Then* rule is commonly adopted in this system for decision making.

Speech Recognition

Speech recognition is one of the most common applications of artificial intelligence in a range of industries, especially online services such as e-Commerce, online learning, language services, and others. This is basically about the capabilities of machines to understand the spoken language and convert it into either text or a speech response to the users. Speech recognition uses natural language processing (NLP) algorithms to accomplish the desired tasks. The most common features and characteristics of speech recognition field are mentioned below [82]:

- The input is taken in the shape of sound vibration and converted into digital signals and then, a range of algorithms are run to understand the meaning of the world and then, apply the software capabilities to convert that word into text or voice.
- This application can perform two types of tasks: Recognition of language words and recognition and differentiation of the voice of a speaker.

- The "Audrey" is the first-ever application developed by Bell Labs in 1952 to recognize a few digits. The other examples of speech recognition include SIRI, ALEXA, Google Assistant, Cortana, and a few others coming out in the recent days.
- Main application of speech recognition apps includes translation, text-to-voice and voice-to-text conversion (transcription), accomplishing voice-based online searches, and other commands issued to machines like drones and others.
- Speech recognition systems or applications can be categorized into two major classes:
  - **Speaker independent systems**—Those that do not need any training to understand and recognize voice inputs is known as speaker independent systems.
  - **Speaker dependent systems**—Those systems that need training through speakers to assess, analyze, and understand the input signals are known as speaker dependent speech recognition systems.
- The most common models, algorithms, and techniques used in speech recognition applications include the following:
  - Natural language processing (NLP)
  - Hidden Markov Models
  - Dynamic-Time-Wrapping (DTW)
  - Deep feedforward and recurrent neural networks
  - Neural networks acoustic modeling
  - N-Gram language model.

Robotics

Robotics is about the creation of mechanical machines that can perform certain tasks, which are already programmed and do not need any human intervention to perform those tasks repetitively. The use of robotics area is expanding in the artificial intelligence field, where the robots can be made more productive, decision-making, and responsive with the help of artificial intelligence (AI)-based software programs, algorithms, and models. Artificial intelligence gives more power, precision, accuracy if it is incorporated properly in the robots. Additional power of AI to robotics increases the scope and roles of robots in a range of fields, applications, and processes across the industries. The future of AI in robotics is very huge, especially in home automation, goods delivery, transportation, military, aerospace, aviation, and other industries [83]. When robotics is associated with communications via the Cyberspace, it can allow amazing things to happen. The advanced e-Healthcare systems for instance are trying to achieve reliable remote surgeries in hospitals, at least at a minor level using robotics, cyber channel and artificial intelligence.

Text Recognition

Text recognition is a field in which the written text in the shape of images such as paper-books, hand-written pieces, and similar types of documents are converted into editable digital text. It is also referred to as Optical Character Recognition (OCR), which is a tool to scan written transcripts and convert into editable digital text.

With the power of artificial intelligence incorporated into the text recognition, its speed, accuracy, and reliability of text conversion have increased significantly. AI-powered OCRs can scan and understand hand-written transcripts, which vary with person to person. A proper training of OCRs through different training datasets can make text recognition a great application. The scope of OCR has also increased significantly with the incorporation of AI in it. It can also check for a range of writing issues made by humans or any traditional text converters. OCRs or text recognition can be divided into the following major categories [84]:

- Intelligent Word Recognition (IWR)
- Optical Mask Recognition (OMR)
- Optical Word Recognition (OWR)
- Optical Character Recognition (OCR)
- Intelligent Character Recognition (ICR).

The most common deep learning (DL) and machine learning (ML) techniques used for making text recognition more prospective include transformers and recurrent layers technologies. The transformers are used similar to the use in natural language processing, while the recurrent layers technologies used in text recognition include:

- Gated Recurrent Unit (GRU)
- Long-Short-Term-Memory (LSTM).

The future of text recognition field powered by artificial intelligence and machine learning is very bright and prospective.

Voice Recognition

Voice recognition is another major field of artificial intelligence in which the components of a particular voice is used to analyze and detect. The application of voice recognition uses voice biometric to recognize a particular voice for any further automatic actions powered by machines such as access control, application logins, and other predefined actions [85].

This is very important to note that speech recognition and voice recognition are two separate domains in terms of their uses. The speech recognition technique is used to analyze, understand, and respond to the voice instructions given through any voice generating application or human. In this technique, the analysis of the words uttered through voice input are analyzed and understood to create a suitable response in the form of either text or a speech created by the machines with the help of a particular software application. At the other hand, the voice recognition does not analyze voice to understand the instructions given through voice but it focuses on the components of the voice such as frequency, pitch, resonance, and others to ascertain that the voice which is analyzed matches with the predefined voice for further actions. This is used like a biometric used in the access of gadgets or other access control systems.

In voice recognition technology, the machines are trained with a range of voice samples for a particular word or phrase uttered in different ways. The machine analyzes and understands the voice with the help of a range of components of that particular voice and saves it in the knowledge-base for matching the incoming commands through voice for future decisions.

Voice-to-Text and Text-to-Voice Conversion

Voice-to-text and vice versa is another important field of application of artificial intelligence in the modern businesses. The core techniques used for voice-to-text and text-to-voice conversion are speech recognition and text recognition respectively. In the voice-to-text application, an input of voice is fed to the machines through internal or external microphone or any other sources. The voice is analyzed and interpreted through predefined algorithms and training knowledge-base developed by the machines. The interpreted language is converted into the text format through other machine-enabled algorithms to write as an output in real-time or recorded formats.

The conversion of text-to-voice is powered by text recognition technique, which takes the input in the form of written or printed text in different forms as discussed in the "text recognition" topic. The text is scanned and processed through machine learning algorithms to generate suitable output in the form of voice. These technologies are extensively referred to as assistive technologies used for learning and other purposes in the field of special education systems and other applications in the field of healthcare, military, and others [86].

Chatbot

Chatbot is another popular application of text recognition applications for establishing a live chatting with the customers or any other end-users in a particular conversion. A Chatbot is a software program that is able to understand the input from the user and respond to them properly to provide the services that he/she requires.

Chatbots can perform a wide range of services in different fields such as customer services, healthcare support, office assistance, web assistance, remote assistance services, service management, product maintenance and management, and many others. Based on the complexity and level of intelligence, chatbots can be classified into three major categories such as [87]:

- Simple chatbots
- Smart chatbots
- Hybrid chatbots.

The simple chatbots work in a very simple way such as an IVR (Interactive Voice Response) applications that we commonly use in telecommunication systems by choosing the right options. This application can easily be incorporated in the form of text while searching for different options in numerous applications. For instance, if we search for some issue in Windows 10 operating system, it will provide you with a range of broad options to choose one from them. As we choose those options, the search is narrowed down to the most relevant issue. Such applications are very simple and based on a predefined fixed algorithm to collect the required information from the users. It may consist of a few instructions and responses.

The other form of Chatbot is known as Smart Chatbot. It is more complex and looks more like-human chatting responder. It has capabilities to understand context, meaning, and sentiments of the users and devise a proper response against the queries based on its capabilities of understanding the conversation through text by the user. The example of such Chatbot include virtual assistant applications, which are used for responding to the queries of clients for a range of issues on a particular online service or web application.

The hybrid Chatbots meet the middle ground between simple Chatbot and Smart Chatbot. These chatbots not only address a wide range of simple applications but also handle the complex conversations simultaneously. The example of such a chatbot is medical diagnosis chatbot, which can handle a range of separate domains of health issues through simple capabilities and then dive into the smart conversation for getting detailed feedback and providing the suitable response simultaneously under the same application environment.

All those types of chatbots use different types of artificial intelligence technologies. A few of those may include:

- Natural language processing
- Natural language understanding
- Natural language generating
- Text recognition
- Speech recognition.

Types of Artificial Intelligence

Artificial intelligence is referred to as capabilities of thinking, sensing, and responding to an action through intelligent computer machines powered by different software and hardware devices. Thus, the baseline for differentiating a type of an artificial intelligence would be the capability and performance of an intelligent system to compare with the human mind. This is the fundamental benchmark to differentiate the types of artificial intelligence in the field of research. At the same time, there are many other criteria on the basis of them—artificial intelligence can also be classified into different types and categories in the field of modern industries and businesses.

Based on the first criteria of the capability and performance of an artificial machine or machine-powered intelligent system, the AI can be classified into four major categories or types, which are [88, 89]:

- Reactive Machines
- Limited Memory
- Theory of Mind
- Self-awareness.

Similarly, based on the most common terminologies extensively used in the business and industries, artificial intelligence is divided into three generalized categories [88, 89]:

- Artificial Narrow Intelligence (ANI)
- Artificial General Intelligence (AGI)
- Artificial Superintelligence (ASI).

The detailed explanations of all the 7 (seven) types of artificial intelligence are described in the subsequent paragraphs.

Reactive Machines

As the name implies, reactive machine is a type of artificial intelligent system that has the capabilities to react against a certain action. In this type of AI machine, there is no capability to understand or learn from the past or other factors around. It can only provide a response to an action based on a set of few pre-programmed instructions. It can provide very limited capability as compared to human mind to respond to any kinds of stimuli in certain conditions. The human response will be different in different stimulus in different conditions; but the reactive machines would repeat the previous action against a particular action or stimulus.

These machines do not have memory-based learning or storing capability to experience from the past and also do not adjust reactions in accordance with any other external or internal learning. These machines are also referred to as primitive artificial intelligent machines, which are designed to provide response to certain actions programmed without any additional capabilities of learning. The example of this type of artificial intelligence is the IBM's Deep Blue machine, which beat Grandmaster Garry Kasparov, world champion of chess game in 1997.

Limited Memory

Limited memory artificial intelligence machines are those machines that have capability of learning from the past experience by scanning, analyzing, and storing the past experience through different types of training datasets in its memory (knowledge-base) in addition to the fundamental capabilities of reactive machine type of AI capabilities. This type of artificial intelligence is also referred to as the Type-II AI in general discourse.

It is very important to note that the "Limited Memory" type of AI is the most mature artificial intelligence that we have in the world today. Almost all applications and intelligence systems developed till today fall in this category of artificial intelligence. The modern AI-powered machines are trained through a range of past experiences or data through video, images, voice, and other types of inputs by building datasets. The machines learn from those training datasets and build a memory-based knowledge-base to use for refining the decision makings. The most common examples of the artificial intelligent systems under this class include:

- Driverless vehicles
- Web chatbots
- Voice chatbots
- Intelligent drones.

Theory of Mind

Theory of Mind is the form of artificial intelligence in which the intelligence machines would learn about the behavior of the users or those that come in contact with them. The behavior learning of those entities would involve a range of factors such as sentiments, body language, psychology, and thoughts of the objects and entities that the machines come in contact. This is the third-generation artificial intelligence, which is under research and development. Until now, no mature machine based on this type of artificial intelligence has been developed for general purposes. Thus, it is imperative to say that all the machines based on AI that we want to build in the next stage fall under this category of artificial intelligence.

As we know, this is a type of AI, which is under research and development; so, more sub-types and categories of this class of AI are likely to emerge in the future for covering a vast area of artificial emotional intelligence. The innovative work on the existing systems would also lead to the development of the intelligent system falling under this category of artificial intelligence.

Self-Awareness

This is the fourth-generation of artificial intelligent system or type of artificial intelligence. This is a conceptual type of AI, which is far away from materialization in the present-day research. Current stage research is still progressing at the 3rd level of artificial intelligence. Therefore, it can be said that the self-awareness type of intelligence is still a futuristic concept.

Self-awareness concept basically deals with the internal presentation of an intelligent system such as internal feelings, capabilities, qualities, behavior, and many other attributes that impact on the interaction with the external world. The understanding and presentation of those attributes for the external entities come after the understanding of the internal behaviors of the external entities as described in the third category of artificial intelligence. In other words, we can say that this type of artificial intelligence deals with the consciousness of an entity. In fact, a human being is very well aware of own behavior, sentiments, senses, and many other consciousness and conscience related factors. By the virtualization of these factors, will it be possible for a machine to interact with its surroundings? Take for instance that a human car driver easily understands that a honking car behind him is governed by the sense of urgency or emergency of that particular driver. This understanding comes in a human-driver due to the understanding of the context, presentation of consciousness or internal behavior of himself/herself. In reality, situational awareness with self-awareness would be indeed very difficult to achieve for machines, even to go close to what the human beings have as their capabilities.

This type of artificial intelligence is the last one that the present-day researchers have visualized conceptually. Before the achievement of this level of artificial intelligence, it is more likely that many other types and domains of artificial intelligence and related technologies may emerge in the future to get this concept materialized at least partially.

Artificial Narrow Intelligence (ANI)

Artificial narrow intelligence, precisely referred to as ANI, is a generalized category of artificial intelligence capabilities in which the machines can perform in line with the 1st and 2nd generation of AI concepts—Reactive machine and limited memory— as mentioned in the earlier topics. This category covers all artificial intelligence-based work completed till today where the machines can perform pre-programmed functions without any additional capabilities of larger decision-making. All machines both simple intelligence machines or limited memory research machines or systems such as automobile cars, computer vision based applications, language processing tools, and others fall under this category of artificial intelligence.

This means, all machines developed till the creation of modern driverless vehicles are the examples of ANI. The earlier machines such as industrial arms for automated tasks programmed through computer for repetitive tasks are also parts of this category of artificial intelligence. As the name implies, the machines or intelligence systems of this category have very limited or narrow range of competency and capabilities. The deep learning, natural language processing, and related processes of AI fall under this category of AI.

Artificial General Intelligence (AGI)

The researchers in the field of artificial intelligence categorized the future work of artificial intelligence that can perform almost similar level of functionalities, capabilities, and competency like a human being does. These machines or intelligence systems would be able to perceive, learn, understand, decide, and function in the similar ways as all those kinds of activities are performed by a human being in our day-to-day life. These systems will be equipped with multi-functional capabilities across all domains of business, industries, and social fields.

Artificial Super Intelligence (ASI)

Artificial super intelligence, precisely known as ASI, is a conceptual category of the artificial intelligent systems in which the machines will surpass the capabilities of human beings in all domains of activities and walks of life. This can also be defined as an age of machines where the machines would be more powerful than human being and even, they become more powerful and out of our control. This condition is referred to as "Singularity" in the field of artificial intelligence. This state is fictitious at this stage of our research on the artificial intelligence. This stage is also known as the pinnacle of artificial intelligence where machines will become a threat to the humans in certain conditions!

This state of artificial intelligence can be justified as an achievable target due to the enhanced capability of processing power of machines, extensively huge memory storage, and extremely greater analysis power. All these factors may contribute to the super learning of the machines, faster processing and analysis of data, and better algorithms to decide in an effective way. The state of AI development till today is far away from such era of super machines. But, there are many researchers who are optimistic to reach this category of artificial intelligence at least up to certain level. At the same time, there are many researchers who are very skeptical about achieving this level of artificial intelligence at any point of time. In reality, the researchers with skepticism should be considered correct because the machines may do many repetitive things better than the human beings (often, even faster) but when human beings act on something, they may also use their emotions on top of the rational thoughts. Hence, we see that sometimes, even if something is rationally correct to do, a human being may not act on that; again, even if something is wrong, a human being willingly may decide to do that. Such human level decision making capability and actions can never be achieved by a machine that is still governed by some set of instructions for its core of the operation, even if this is called some kind of 'learning' or 'self-learning'!

Intelligent Agent and Environment

An artificial intelligence system conceptually consists of two components that interact with each other to accomplish the entire process of understanding and decision-making for a required action to be taken by an intelligent machine or system. Those two components include:

- Intelligent Agent (AI)
- Intelligent Environment (IE).

The environments are the surroundings of an intelligent system used for taking or inputting the raw data for intelligent agents as well as sending (output) the actions performed by the agents into it. The major function of agent is to take input from the environment through a range of sensors for developing perception in the intelligent system and also sending the action after building perception through actuators to the environments. This entire process is supposed to be automated (powered) by the intelligence of the system, mostly generated by the intelligent agent. The schematic diagram of the environment and intelligent agent is shown in Fig. 2.2.

Both components play pivotal roles in understanding the concept of artificial intelligence. Let us now explore both of them separately at length.

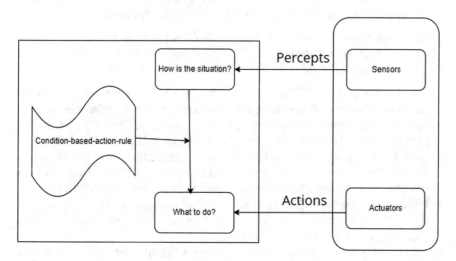

Fig. 2.2 Schematic diagram of intelligent agent and environment

Intelligent (or, Intelligence) Agent (IA)

Intelligent Agent is an entity in an artificial intelligence that acts as decision maker through perception function by taking input from the environment through a sensor and acts through actuators in the AI environment. In other words, intelligent agent, precisely IA, is an entity that takes decisions and initiates the action in an artificial intelligent ecosystem. Thus, an intelligent agent performs two functions—building perception or understanding of the environment conditions and initiating action after processing the input information through certain pre-defined algorithms [90].

There are two major types of intelligent agents (IAs) commonly used in a hierarchical structure of intelligent agents. They include:

- **Lower-level agents**—These agents perform low-level tasks. They are also referred to as sub-agents in the hierarchical structure of an agent system.
- **Higher-level agents**—These agents perform high-level tasks in the agent hierarchical structure.

As mentioned here, the structure of an intelligent agent consists of three major components as listed blew:

- **Architecture**—This refers to the hardware devices or machines on which the intelligent agent works through sensors and actuators. For example, a motor car, personal computer, and others.
- **Agent program**—It is a programmable software or codable program that resides on the hardware and performs the core functions of an intelligent agent in an intelligent machine or system.
- **Agent function**—This refers to the mapping of actions that are the outcome of the perception sequences on which the actions are mapped. The perception sequences are generated by the intelligent agent through historic learning.

Intelligent agents are categorized on the basis of their capabilities and level of perception building or understanding from the environmental input. There are five (5) major categories classified on these criteria as mentioned below:

- **Simple reflex agent**—Those agents that take action on the basis of current percepts rather than historic perception are known as simple reflex agents.
- **Model-based reflex agent**—An agent that works on the basis of history perceptions to take action is known as model-based reflex agent.
- **Goal-based agents**—This is relatively more advanced than the above two agents because it possesses higher capabilities to from a range of actions based the main objective of achieving the desired goal specified for a particular function.
- **Utility-based agents**—Based a utility function, this agent makes decision in the light of the utilities of the intelligent agent for which it is designed. It is more capable agent than the previous three agents.
- **Learning agents**—This is the most advanced intelligent agent, which is capable of learning from the environment. It consists of the following components/elements that make it more advanced and more capable of all other agents:

- Learning element
- Performance element
- Critic element
- Problem generator.

Let us summarize the major characteristics of intelligent agents (IA) based on the above discussion and its additional features:

- They possess learning capabilities through percepts
- They are autonomous based on the action-rule to take action
- They have capability of interacting with other agents, human, machines, and systems
- Ability to accommodate new rules through incremental procedure
- Intelligent agents are knowledge-based and show goal-oriented habits.

Artificial Intelligence Environments

An environment is the surroundings of an intelligent agent in an artificial intelligence ecosystem. An IA takes input from the environment for perceiving the situation of the environment through sensors and returns with an action on the environment based on the decision made through a range of action-based rules. There are numerous types of artificial intelligent environments used in the modern works of AI research. A few of them include [91]:

- **Fully-observable versus Partially-observable**—A type of environment in which the sensor of an intelligent agent (IA) can see entire environment at any point of time is called fully-observable environment. If the agent has limited access to the environment through the sensors, it is called partially-observable environment. Again, if there is no sensor for an IA to sense the environment, it is called as non-observable environment. Observing chess board through camera is the example of fully-observable environment because you can see both pawns and moves fully. Front-camera in a driving car is an example of partially-observable environment.
- **Single-agent versus Multi-agent**—As the names imply, the single AI environment is meant for a single intelligent agent while the multi-agent environment is developed for multiple agents in it to act.
- **Deterministic versus Stochastic**—The environment in which the uniqueness in current state of the agent can be determined like, what are the next possible options to take, is known as deterministic environment. For instance, an environment of a chess board where on the basis of the current state of a pawn, the next possible moves can be easily be determined. Such environment is known as deterministic environment. On the other hand, the stochastic environment is that in which the current state of the agent does not help in determining the possible future options. Those moves would be based on numerous variable factors, which will emerge with every changing fraction of time. Such environment is known as stochastic environment of artificial intelligence.

- **Competitive versus Collaborative**—The competitive type of environment is that in which two or more AI agents act to win over the other by using the emerging situations and conditions in the best way. The example of competitive environment is the chess game in which two agents try to win over the opponent by using the best available intelligence in the environment. Again, the collaborative environment is that in which the intelligence agents collaborate with the other agents in the environment to make the system work smoothly. An example of this could be multiple driverless cars running on a road in a collaborative environment to avoid any mishap and collisions.
- **Episodic versus Sequential**—Episodic is the type of AI environment in which an agent takes action in isolation from the previous action as well as the future actions. For instance, trash picking artificial agent will pick the track wherever it sees. It does not relate its action with the previous one or the future one. It acts from scratch after every action is completed. On the other hand, the sequential environment is that AI type of environment in which the present action is dependent on the previous actions taken or the future actions to be taken.
- **Static versus Dynamic**—The static environment is an AI environment, which does not change with the actions of AI agents or time or even other parameters. The example of such environment is an office premise which does not change if a robotic AI agent cleans the entire office and then serves a cup of tea to an employee. On the contrary, the dynamic environment changes with the time and actions of AI agent. The roller coaster simulation changes with the time and actions after every point of time and distance. Such kind of environments are known as dynamic artificial intelligence environments.
- **Discrete versus Continuous**—The type of artificial intelligent environment in which a definite number of actions are performed in an order to achieve final goal of the application is called the discrete environment. For example, a definite number of options are available for taking one move which affects the next move too. Thus, a number of functions are to be performed to win the chess game. On the other hand, the continuous environment is that one in which an AI agent has to perform unlimited (indefinite) number of actions in a continuous manner. The driverless car AI environment is a good example of continuous AI environment in which AI agent performs indefinite number of actions such as parking, moves, breaks, and others.

Future of Artificial Intelligence

The future of artificial intelligence is very bright, promising, and impactful on numerous domains, industries, human behaviors, social livings, governments, and global communication. AI has started passing through the second generation as envisioned by the technology experts and business gurus. Dr. Kai-Fu Lee once expressed his expert views in 2018 that, "*Artificial intelligence is going to bring changes in the entire world more than anything in the whole history of mankind*".

It is imperative to know that artificial intelligence started long before, in the 1950s and 1960s. But, it went through numerous roller coaster rides for many times. Initially, it promised to change things abruptly but even after many research works and publications, the actual impact was quite low for some decades. Slowly and gradually, it started gaining grounds in the 1990s and 2000s. The uptrend of the growth of artificial intelligence and impact on the human lives as well as businesses have already started getting noticed by this time. Still, the maturity of artificial intelligence is yet to be achieved to the expected level.

If we consider the third and fourth generations (periods) of artificial intelligence, it will take a long course to achieve those objectives. But, one thing is very sure and that is, the impact of AI will remain very high on all domains of industries, society, government, and global relationships. The impact of AI will bring forth many new domains of knowledge, areas of opportunities, types of industries, and a wide range of challenges in the future too. The main areas that boosted the growth of artificial intelligence include the following [110–112]:

- Software development technologies, programming languages, development platforms, and methodologies.
- Extensive growth in the speed of the computing resources or computer processors. According to the Moore's Law, the computer processing power or speed is doubling after every two years and the cost of the machines is continuously dropping significantly. This is because of the fact that the number of transistors on the microchip is doubling that reduced the cost by nearly half the previous cost.
- High speech Internet is another giant contributor to the growth of artificial intelligence by providing an opportunity to integrate with multiple sensors and remote locations and applications to generate a desirable impact on the AI field.
- **Internet of things (IoT):** This field came into existence due to effective integration of numerous communication technologies and sensor technologies, which provide a vast field for the artificial intelligence technologies to grow.
- Extensive growth in the neural networks for deep learning and machine learning technologies powered by a range of software-based applications and fields of intelligent machines such as computer vision (CV), data analytics, robotics, process automation, and many others, which laid the foundation for the futuristic building of AI field.

With the advent of modern and highly sophisticated software programming concepts, the new domains got materialized which are extensively driven by the artificial intelligence and machine learning. The industries and technological fields have emerged due to the desirable impact of the artificial intelligence and machine learning. A few of them include the following [110–112]:

- **Computer vision (CV):** The computer vision technology paved the way for supervised and unsupervised types of machine learning for the artificial intelligence systems. Thus, the technology started evolving with the power of artificial intelligence.

- **Big data**: Big data management became possible through numerous repetitive processes managed by the automated functions and processes powered by artificial intelligence and machine learning.
- **Data analytics**: Artificial intelligence impacted this domain hugely due to the availability of software applications that use machine learning and deep learning at the core of their code, which can analyze a huge volume of raw and unstructured data within a short period of time with greater accuracy to provide one with a deeper insight into the business intelligence (BI) hidden in that raw data.
- **Robotics**: The most common advancements in the modern robotics powered by the artificial intelligence include unmanned aerial vehicles (UAV), commercial drones, industrial automation, home automation, office automation, and many others.

There are many general functions, activities, and processes that have drastically been changed due to the pervasive impact of artificial intelligence and its sub-domain, machine learning. A few of those impacts are listed below [110–112]:

- The power of visual inference has increased significantly due to the impact of AI applications. For example, we infer 25 signals in an image today, it can be more than 100–150 inferences of signals from the same image a few weeks later with the help of modern AI-powered software applications.
- Very high deployment of artificial intelligence and robotic applications have revolutionized the modern healthcare, especially the medical surgeries, healthcare management, remote healthcare support and other domains. The influence will continue to increase in the future in all domains of healthcare.
- Driverless cars or automobiles are another master piece of artificial intelligence that will take over the traditional systems of transportation with new norms and new traditions. This can also leave millions of vehicle drivers jobless.
- Our privacy and human rights can get affected badly if artificial intelligence is not governed through proper regulations and laws.

Let us now explore the main contributing factors in the modern AI market, which is noticing amazing growth and prospects in all domains of industries and societies. According to Statista projections, the global market size of the revenue of only artificial intelligence software is expected to cross 126 billion USD by 2025, which was just 10 billion USD in 2018 [113]. The total investment in research and development in the field of AI was over 20 billion USDs in 2021. This huge investment is mainly dominated by the big technical giants such as Google, IBM, Microsoft, Oracle, Cisco, and so on.

All jobs and professions are going to be influenced heavily with the growth of artificial intelligence. All major jobs will remain fully dependent on the computer coding or programming. A doctor will not be able to work efficiently without the expertise of computer coding to deal with the modern robotics and AI-powered machines (some trends are already being seen). All routine and repetitive activities of maintenance and operations that are well-scripted will vanish or will be performed by the robots that will be maintained under highly skilled and knowledgeable engineer or programmer. Well, we can at least be optimistic! But, human beings may not be eventually deducted from the scene of controlling all these machines!

Another impact of artificial intelligence will emerge through the inventions of the following very critical technologies in defense and military fields:

- Development of robotic soldiers for hybrid wars
- Development of swarm drones with full coordinated communication without any external support from humans
- AI-powered nefarious designs adopted by the hackers and other malicious actors that can unleash huge damage on the society.

The impact of these adverse things for humanity may lead to devastating impact if not controlled properly. For example, the development of modern robotic soldiers would lead to jobless human-soldiers who have at least some feelings and emotions. The robotic soldiers would be only machines and the targets will remain only targets without any differentiation. The swarm drone is another form of automated soldiers or drone-based weapon systems which will communicate and coordinate with each other to destroy the targets in a very small size of them with a huge swarm of those drones.

The School of Humanity of Oxford University published the results of a survey regarding the capabilities of robotic automation and human standing in the future. Those results are astonishing. Here, we have summarized them for the future impact of artificial intelligence:

- Machine will become capable of writing essays by 2026
- By 2027, the drivers will be rendered unnecessary roles by driverless vehicles
- AI will outperform human in retail field by 2031
- AI will become the next Stephen King by 2049
- AI will become the next Charlie Teo by 2053
- AI will automate all human jobs by 2137.

Sample Questions and Answers

Q1. What is the Symbolic approach of machine learning? Explain.

A1. Symbolic approach is also referred to as Symbolic Artificial Intelligence, precisely SAI. The symbolic approach is based on the notion that the whole world can be taught to the machines through the input of symbols used to denote an entity in the world. The symbols are the fundamental units or building blocks of the entire world. The operations between the symbols are considered for making the machine algorithms understand the response to the situation when a query is put up by a user or automated system connected to the artificial intelligence-based machine.

Q2. What do you mean by *learning* in Artificial Intelligence?

A2. Learning is one of the most basic goals of Artificial Intelligence. It is a process, which enables the machines to understand about the environment, which includes a range of information such as voice, objects, motion, blow, winds, sun, heat, temperature, and many other things. This process is similar to the capability of a human being to learn naturally about the environments that he/she comes in contact with.

Q3. What is machine perception? How can it be built?

A3. Building machine perception refers to developing the capabilities in machines to sense, see, listen, taste, touch, and feel through the processing of different types of signals originated through different sensors. A range of metrics are used for training the machines to build perceptions regarding the above-mentioned senses. By developing such capabilities in machines, numerous applications can be built to automate a wide range of processes, activities, and functions easily.

Q4. Name the two major categories of speech recognition system.

A4. Speech recognition systems or applications can be categorized into two major classes:

- **Speaker independent systems**—Those that do not need any training to understand and recognize voice inputs is known as speaker independent systems.
- **Speaker dependent systems**—Those systems that need training through speakers to assess, analyze, and understand the input signals are known as speaker dependent speech recognition systems.

Q5. Can artificial super intelligence be really achieved? Support your answer with appropriate rationales.

A5. Artificial Super Intelligence (ASI) can be taken as a goal in the AI research; however, the fact of the matter is that the machines may do many repetitive things better than the human beings (often, even faster) but when human beings act on something, they may also use their emotions on top of the rational thoughts. Hence, we see that sometimes, even if something is rationally correct to do, a human being may not act on that; again, even if something is wrong, a human being willingly may decide to do that. Such human level decision making capability and actions can never be achieved by a machine that is still governed by some set of instructions for its core of the operation, even if this is called some kind of 'learning' or 'self-learning'! However, as mentioned, as a research objective, ASI can push the researchers to work further in this direction.

Test Questions

1. What is Big Data and how does it work?
2. In what sense computer vision is used?

3. How did artificial intelligence grow?
4. Is there an environment that enables Artificial Intelligence (AI)?
5. How do you define Artificial Narrow Intelligence (ANI)?
6. What are the types of Artificial Intelligence?
7. How does Optical Character Recognition (OCR) work and what are some of its categories?
8. Why is speech recognition important?
9. Is robotics a science? Justify your answer.
10. How does Natural Language Processing (NLP) work and what are its major components?
11. Why do we need business intelligence?
12. How does Artificial Intelligence work?

Chapter 3
Machine Learning Technology

Introduction to Machine Learning

Machine learning, precisely referred to as ML, is a sub-domain of artificial intelligence (AI) technology. Artificial Intelligence is a large field dealing with the computer science and mathematics for materializing numerous solutions to the problems through machine intelligence or machine's capabilities to deal with the problems like the human-being does. Thus, machine leaning is also a field that deals with computer science, mathematical analysis, mostly statistics, and data science/analytics. The power of machine learning technologies improved significantly with the advent of modern computer programming or software development. Thus, machine learning can be defined as [115]:

> A branch of artificial intelligence (AI) technology that uses the structured/unstructured raw data and mathematical algorithms powered by the software programs, which enable the machines to imitate the ways humans learn from the real-world environment and improve accuracy with the increased experience with the real-world data without being explicitly programmed through computer instructions to do so.

From the above-mentioned definition, it is very clear that the combination of input of data from the real-world environment in the form of training datasets—either labeled or unlabeled—are fed to the machines as an input, which is processed through internal algorithms. Those algorithms are logical frameworks based on mathematical analysis and coded in the software programs, which imitate like the logical thinking of a human brain. Those programs take the input data as consumables and process it to learn and understand from that data by imitating like a human does. The most important part of this definition is that the machine learning systems are not explicitly programmed, which means that they are trained to learn from the raw data and understand the way human does through following the algorithm process and make decision for any incoming scenario in the light of previous learnings that the system has achieved not on the basis of any explicit program to follow to make decision. Thus, it is called as "*without explicit program*". Though, the algorithms are input output sources and other training components are programmed through computer programming instructions, the decision making after learning the environment is made purely on the basis of the learning of the intelligent machines or intelligent systems.

Diving deeper, let us consider a few common terminologies that are mostly used interchangeably for machine learning (ML) or artificial intelligence (AI) by many people and also create certain levels of confusion in the minds of the readers, such as:

- Deep Machine Learning (DML)
- Artificial Neural Networks (ANN).

Deep Machine Learning, Machine Learning, and Artificial Neural Networks are all sub-classes of Artificial Intelligence (AI); thus, the mother of all these three terms is AI. Machine learning is a field of artificial intelligence that uses labeled training datasets to learn about the environments and scenarios to decide like a human. On the other hand, deep machine learning is a field of artificial intelligence that uses raw or unstructured data without any labeling or annotation to learn by experience. The deep learning deals with the feature extraction processes' automation to enable the machines to build and utilize larger datasets without any major role of human intervention. Thus, it can also be termed as scalable machine learning [116].

The artificial neural networks (ANN) are the network of layered nodes consisting of input node, output node and hidden layers. The hidden layers can be one or more than one consisting of a series of nodes to decide about the logical process steps.

Each node in the neural network has a threshold value and associated weightage. If the value of a node reaches the threshold value; it gets activated and signal is processed further based on a logical configuration. The weightage of that node is considered in the artificial learning process. Neural networks also define the depth of the learning. If the hidden layer consists of more than one layer; those networks are considered as deep learning neural networks depending the level of depth on the number of hidden layers. If the hidden layer consists of just one layer and total neural network including input layer and output layer consists of three nodes; it is simple or basic artificial neural network.

Importance of Machine Learning in Modern World

According to the Deloitte Survey 2020, more than 67% of the companies surveyed were already using machine learning in their respective business processes and more than 30% more companies were planning to use the same in their respective business processes by 2021 [117]. The total size of machine learning worldwide is expected to reach $152.24 billion by 2028 from just $11.33 billion in 2020 with a whopping growth of 38.6% CAGR (compound annual growth rate) during the forecast period [118]. If we see the retro-market size of ML, the total global market size in 2017 was just $1.4 billion. On the basis of such promising statistics, it is very easy to estimate the importance of machine leaning in the modern world of business across all parts of the world.

The most common areas of businesses, where the impact of machine learning has been seen tremendously include the following [116]:

- **e-Commerce**—In e-Commerce, a range of activities and processes have been automated with the help of machine learning incorporated through software programs such as customer interest analysis, product recommendation, cross-selling support, automated real-time help, order form filling, computer vision based product checking, and many others.
- **Customer support**—Customer support is one of the most important fields where the impact of machine leaning is seen huge. The examples of usages include chatbots, real-time virtual assistance, technical support guides, and many others.
- **Online trading**—The automated trading online in different markets such as stocks, currencies, and commodities uses the power of machine learning. Automated trading performs millions of transactions without any intervention from human-being.
- **Language services**—A huge ratio of many language services and support is using the power of machine learning's sub-domain named as natural language processing (NLP). The translation of one language to another one is being performed by the automated systems, text is being converted to voice and vice versa with the help of machine learning-powered tools. Text checking and editing

is another major domain which is being taken over by the ML-powered systems gradually.

- **Entertainment and media**—All major entertainment applications such as gaming, social networking, videos, and other entertainment as well as news media are using computer vision based applications tremendously. The use of CV (Computer Vision) technologies has changed the face of entertainment with new looks via virtual ecosystems.
- **Marketing and Business Intelligence (BI)**—Modern businesses have become so competitive and knowledge-driven that any decision that digresses from the main target loses its direction heavily. Machine learning helps businesses dig into the heaps of raw data to find out the valuable business intelligence to make data-driven decisions in marketing, business strategies, and business management.
- **Miscellaneous**—Almost all types of industries and businesses are using the power of machine learning in one way or the other. The most common of those industries include software development, web hosting, telecommunication applications, aviation, defense, sales, data analytics, research and education, and so on.

How Does Machine Learning Work?

Before diving into the working principles and procedures of machine learning technology, let us describe the main features of the work performed by it. The most common features of a machine learning activity of function can be divided into the following three categories:

- **Descriptive function**—This function of machine learning system uses the data to learn and explain what has happened in the given situation through a descriptive capability. This feature helps figure out the details of the activity or incident.
- **Perceptive function**—This capability of machine learning enables the users to get suggestions through the analysis of the data from the machine learning systems about what action is to take in the given situation of the data.
- **Predictive function**—This feature enables the intelligent systems to project about the future situation. What will happen under the given conditions based on the raw data? Intelligent systems can help through predictive capability.

Before going into the technical jargon of the working process of machine learning, let us know a bit more about the working of machine learning systems in simple words. At the first step, the raw data in the form of labeled or unlabeled training datasets is fed to the ML algorithm, which takes the data into the system and makes an assessment or estimation of the situation. This is called learning process. After the system has made estimates, with new data, it is checked if the output is as per desired outcome in the given situation with the help of the understanding of the machine learning system that it had achieved in the first phase. If the result is not as per desired outcome, the system is trained further to make it more accurate. The second phase is known as testing phase in general terminology. The last phase is verification of the data phase—it is when the input data is fed and the system generates the desired outcome.

Technically, this entire flow can be divided into three major phases as mentioned in the following points sequentially [116, 119]:

- **Decision process**—In this phase, the data is fed to the ML systems in the form of training datasets—either in labeled or unlabeled forms. The machine learning system will generate an estimation pattern, which will be used for the decision making when tested with the further data input scenario.
- **Error function process**—In the second phase of the machine learning flow, the ML systems will assess the error in the function or outcome if the previous data exists. This phase checks the previous understanding of the system in predicting the output in real-world data scenarios. If the system generates erroneous results, the difference between the desired results and erroneous result are compared to generate an error function.
- **Model optimization process**—This is the last phase of machine learning that continues as the database or knowledgebase of the system increases. In this process, the ML model is tested and verified by providing more data through training datasets to make the prediction more accurate and reliable. It is tested, repeated, and verified till the desired objectives have been achieved through the ML project.

Thus, the entire working principle of ML models is based on feeding raw data for learning and building an *estimate model* and then, checking that estimate model through real-world data of the same scenario. If the estimate is inaccurate or away from the desired objectives, the system is trained further to make it more mature and again tested for its results. This process continues in the third and last phase until the model produces the desirable accurate and reliable results.

Types of Machine Learning

Machine learning technology is divided into sub-categories on the basis of the level of human intervention to achieve goals of intelligent systems. The most common types used in the modern machine learning technology include:

- Supervised Machine Learning
- Unsupervised Machine Learning
- Semi-Supervised Machine Learning
- Reinforcement Machine Learning.

The details of all of the above-mentioned four major types of machine learning are given in the following topics respectively.

Supervised Machine Learning

Supervised machine learning, also referred as supervised learning, is a type of machine learning in which the training datasets are created by labeling and annotation to feed to the algorithm for developing understanding of the data by major functions such as classification and prediction. The classification function identifies the possible category of the input items and the prediction functions draws results as accurate as possible. The most common differentiator of supervised machine learning is that it needs manual human intervention in training the machines with the real-world environment by tagging and annotating the items in different types of data such as text, images, videos, audios, and a range of sensor data in different forms and formats [120].

This entire process of supervised machine learning consists of a range of algorithms that are designed to solve the problems related to the following main categories:

- Classification
- Regression.

These two domains use different kinds of algorithms. A few very important algorithms used in classification problem solution categories include:

- Linear classifier algorithm
- Decision tree algorithm
- Support vector machines (SVM)
- Random forest algorithm
- K-nearest neighbor algorithm.

On the other hand, the main supervised machine learning algorithms used in the category of regression problem solutions include:

- Linear regression algorithm
- Polynomial regression algorithm
- Logistical regression algorithm
- Lasso regression algorithm
- Ridge regression algorithm.

The most common examples of the use-cases of supervised machine learning include predictive analytics, customer behavior and sentiment analysis, object recognition, movement recognition, face recognition, spam detection, and so on.

Unsupervised Machine Learning

Unsupervised machine learning, which is precisely referred to as unsupervised learning, is the second very important type of machine learning. In this type of machine learning, the unlabeled or untagged datasets are used for training the machine learning algorithms to conduct cluster analysis on those datasets. The algorithms used in unsupervised learning discover range of features and patterns without any manual human intervention [121].

Unsupervised machine learning does not need any labeling, tagging, or annotation of data to be used for building training datasets. Unsupervised machine learning algorithms are designed for carrying out three most common functions as listed below:

- Clustering analysis
- Association establishment
- Dimensionality reduction.

The clustering is a process of artificial intelligence in which algorithms group different items and data components in an untagged training dataset based on the similarities and differences. There are different types of clustering analysis and they use different algorithms for their respective functionalities as mentioned below:

- **Overlapping and Exclusive clustering**—The overlapping clustering uses algorithms that analyze the degrees of membership in two separate groups. The example of overlapping clustering algorithm includes Fuzzy K-means. On the other hand, exclusive clustering, also referred to as hard clustering, analyzes the uniqueness of the features to be part of a group. The main example algorithm of this task includes Hard K-means algorithm.
- **Hierarchical clustering**—The algorithm used in hierarchical clustering analysis include Euclidean distance algorithm, which may be used in two different approaches commonly referred to as bottom-up and top-down approaches.
- **Probabilistic clustering**—This is the clustering analysis that uses the laws of probabilities to group an item in a certain category. The most commonly used algorithm for this task is Gaussian Mixture Model (GMM).

The second major function of unsupervised machine learning systems is to establish association or relationship among the variable in untagged datasets. This task is performed through different algorithms. One of the most popular algorithms used for association rules is Apriori algorithm. The establishment of association of variables is based on certain rules. The other algorithms used for association rules include Eclat and FP-growth algorithms.

The third task of unsupervised machine learning models is dimensionality reduction, which is used to strike off the most irrelevant or negligible values of different features or dimensions of datasets. These are reduced to generate manageable results but the accuracy is not impacted significantly. As the supervised training grows bigger, a huge range of features and dimensions of dataset variables are created, which generate more accurate and reliable results but they also result in low performance of the algorithms. In such conditions, it is good idea to ignore those features that have minimum or negligible impact on the accuracy of the results so that a management outcome of unsupervised machine learning can be achieved. A few very important methods used for dimensionality reduction include the following:

- Principal component analysis (PCA)
- Singular value decomposition (SVD)
- Autoencoders.

The most common examples of use cases of unsupervised machine learning include computer vision applications, medical imaging processes, news article categorization, customer persona, recommendation engines, anomaly detection, and so on.

Semi-supervised Machine Learning

Semi-supervised machine learning is a hybrid type of machine learning that uses features and capabilities of both supervised and unsupervised machine learning. In other words, it is the middle ground between those two types. In this kind of machine learning, a small volume of labeled data is used to train the ML-model first and then a huge volume of untagged or labeled data are used for training the machine learning models. Then, some assumptions are used to merge the learning output to obtain the final outcome. The entire working process can be explained with the following steps [122, 123]:

- First of all, a small volume of labeled training datasets is fed into the algorithm to learn about the input data or real-world information. This entire process involves all activities such as testing training, and verification of the training to find out the capability of the model to produce the desired outcome.
- In the second step, a large volume of pseudo labeled data is fed to the algorithm for training purpose. These unsupervised training datasets will not yield accurate results, though.
- After feeding both training datasets—labeled and unlabeled, the tags or labels and pseudo labels are linked for establishing association among them.
- In the fourth step, the labeled and unlabeled input training datasets are linked with certain assumptions and rules, governing the entire linking process.
- To achieve more accuracy, the first step of supervised learning is repeated to increase the reliability of the outcome of this model.

The main reason to use a hybrid model of machine learning is that the supervised machine learning is very costly and time consuming due to the requirement of human intervention through manual tagging and labeling. The quality of tagging is also one of the main concerns in the supervised learning. Thus, it becomes very unviable for the modern businesses to use this method only. Meanwhile, the unsupervised training scope is limited, which means there are a very few scenarios or conditions where the unsupervised machine learning can be used to achieve the desired objective. Thus, the best trade off out of these conditions is the use of hybrid machine learning algorithm.

A semi-supervised algorithm makes certain assumptions for the classification, clustering and other operations as mentioned below:

- **Continuity assumption**—The data points which are closer to each other are more likely to have same output labels.
- **Cluster assumption**—Data can be separated into discrete clusters. The points in the same cluster are more likely to have the same output data label.
- **Manifold assumption**—Normally, data lies on manifold of lower dimension than space. Thus, the distances and densities can be used.

The most common real-world business application of semi-supervised machine learning are mentioned below:

- Classification of protein sequencing in medical science
- Text document classifier applications or tools
- Speech analysis and recognition tools
- Classification of online web content.

Reinforcement Machine Learning

Reinforcement is one of the three major types of machine learning. In this method, training to the ML-model is done through feedback of the step taken by the AI-agent in the AI-environment. If it takes the right step, a positive feedback or reward is received and if it takes a wrong step, a negative feedback or penalty is received by the ML-model. In this training, the AI-agent learns by itself by taking suitable actions in the given environment. Reinforcement machine learning can be described through some features and characteristics such as [124, 125]:

- It is a type of self-learning model for an intelligent agent about the environment through actions and its results—positive or negative.
- It does not use labeled data for learning purposes rather learns through self-experience of interactions with the components in the AI environments.
- It is a type of machine learning where the decision making is sequential depending on the output of the step or action taken in the environment.
- The objective of this type of machine learning is long-term projects of machine learning through real-world experience.
- The most basic role of AI agent in reinforcement learning is to improve the performance of the machine learning model by achieving the maximum positive score through rewards.
- This model is also known as hit and try model of machine learning.
- Reinforcement machine learning plays very pivotal role in the field of artificial intelligence field where the AI agents are commonly deployed for reinforcement learning.

- There is no guidance for an AI agent about what to do and there is also no introduction of the environment to the agent. It is all about hit and try to learn about the situation around to achieve the most suitable solution.
- The next action of an agent always depends upon the feedback of the previous step and it changes the state accordingly.
- The environment type used in the reinforcement machine learning is stochastic-environment.
- The work-flow of reinforcement learning consists of:

  – Agent takes an action on the environment by its own
  – Environment provides the feedback—positive or negative
  – Agent either changes state or maintains status quo and takes the next action based on the previous feedback
  – Again, the same process will repeat and the outcome value is calculated on the basis of Bellman Equation

- Reinforcement machine learning uses Bellman Equation, which is mentioned below:

$$V(s) = Max\ [R\ (s,\ a) + \gamma\ V(s')]$$

where

V(s)   **The total value calculated at a particular point**
R(s, a)   **The reward by an action "a" at a particular state "s"**
γ   **Discount factor (Gamma)**
V(s')   **Total value at previous state.**

- There are two major types of reinforcement machine learning type, which are:

  – **Positive reinforcement**—Adding value to increase the tendency, which is the expected behavior of an agent that would occur again as a response to a given action.
  – **Negative reinforcement**—Adds a value to reduce the tendency to the behavior of agent to occur again to avoid negative feedback.

- There are four main components of reinforcement learning that include:

  – **Policy**—It is the fundamental component, which defines the behavior of an agent in the given AI environments
  – **Reward Signal**—The reward signal is an indicator of feedback to the intelligent agent after it takes an action on the environment
  – **Value Function**—This value defines both the state and actions that are good for the future actions
  – **Environment Model**—This is the environment simulation, which helps estimate the behavior of an environment in the light of states and actions

- Reinforcement machine learning uses different types of machine learning approaches in the ML-models. Top three of them are:
  - Policy-based approach
  - Model-based approach
  - Value-based approach

- The most important use cases of reinforcement learning in the modern machine learning and artificial intelligence fields include robotics, control system, gamming, chemical reaction optimizers, industrial automation applications, building strategies for different business processes, and many others.

What Is Deep Machine Learning?

Deep machine learning, precisely known as DL, is a type of machine learning, which is powered by the use of neural networks to learn about the problem without any human intervention. Deep learning does not rely on the manually tagged data but it uses the power of neural networks, which play the role like a human brain to learn about the problems deeply through its nodal structure. It uses huge data for learning about the problems [126]. In simple words, when usual machine learning relies on fixed set of algorithms, the deep learning (DL) structures algorithms in layers to create an "*artificial neural network*" that can learn and make intelligent decisions on its own, i.e., it may rather try to create own algorithm to use for learning and decision making (at least, that is the concept).

Artificial Neural Network

An artificial neural network, precisely referred to as ANN, is a structure of multiple layers for processing the data to develop a deeper understanding of the problem without any human intervention. A neural network has at least three types of layers such as input layer, output layer, and hidden layer. Each of the layers may comprise of multiple nodes. The hidden layer may consist of multiple layers made up of numerous interconnected nodes. The nodes in those layers are known as neurons, which work in the similar ways as the human neurons do. Each node/neuron has a threshold value and associated weightage of the node. If the input signal value is less than that threshold value, the node does not get activated to pass the signal to the next layer nodes. If the input signal crosses the threshold value of a node, it will get activated and will pass the signal to the next layer connected to it. Thus, the overall value of the process learning is calculated through related mathematical expression.

Major Methods/Techniques of Machine Learning

Machine learning is a very important domain of artificial intelligence. It plays very vital role in the modern process automation and robotics. Machine learning is continuously expanding its volume by adopting new models, methods, and techniques to provide solutions to a wide range of issues in the modern science and technological spheres. Among those techniques and methods, a few very important ones, which are extensively used in modern machine learning technology are explained below [127].

Regression Model

The prediction of numbers or numeric values as an output or result is achieved through a machine learning model, which is known as regression technique. In this technique, the output is a continuous or real number. This technique is extensively used for the prediction of numeric values in the output of a problem through a range of algorithms based on different mathematical expression, rules, and formulas, which will be covered separately. The most common use cases of this technique in the modern industries include demand, supply, market growth and many other similar kinds of predictions that involve numeric values as the results.

Decision Trees

In decision tree technique, the machine learning algorithms classify and estimate the answer based on their features in flow-chart based on the branches of a tree. The features of internal node presented in the tree-branch structure reaches the leaf node eventually. The leaf node's characteristic is the resultant class based on the features classified through decision tree technique. This is a very easy and effective technique adopted in the machine learning systems.

Clustering

Clustering is the process of grouping the data points based on the characteristics that relate to the resemblance or the unfamiliarity with the other data points. Clustering is extensively used in the supervised machine learning for categorizing different data points based on their characteristics. This is a type of statistical approach for analyzing and grouping the data points or variables. Numerous algorithms are adopted for achieving effective clustering or cluster data analysis in the modern projects of machine learning and artificial intelligence. More details and explanations of those algorithms will follow in this chapter later.

Classification

Classification is another very important technique used in machine learning field. This technique is used for supervised machine learning for classifying a data point into one or more classes. It can produce output of one or more than one value based on different classifier algorithms, which will be described separately later in this chapter.

Anomaly Detection

Anomaly detection technique-based algorithms find out the data points that are different from the other data sets. The anomaly detection model deals with the outlier data that is way too (or significantly) different from other data sets under consideration in a particular machine learning model. Those points are also referred to as the unusual data points. The detection of such points is done in three major types of machine learning such as supervised, unsupervised, and semi-supervised models. The most important use cases of anomaly detection models in modern industries include cybersecurity in banking frauds, network operations, and so on.

Neural Network Method

The neural network model uses many mathematical techniques and formulas in different algorithms to learn the data in much deeper way. The details of artificial intelligent neural networks have been provided in this chapter.

Dimensionality Reduction

Dimensionality reduction is the method of simplifying numerous redundant, sub-category, irrelevant, and duplicate characteristics or features of a data point. This is used for predicting the class of that particular data point. The dimensionality reduction technique is extensively used in the machine learning problems related to classification of data points based on a huge number of characteristics or features. In fact, use of a huge number of variables (features) to predict and classify a data point makes it very difficult to work within a model while maintaining the performance of the system. Dimensionality reduction uses various algorithms to achieve the desired objectives such as [128, 129]:

- Principal component analysis (PCA)
- Linear discriminant analysis (LDA)
- Generalized discriminant analysis (GDA)
- And others.

The above-mentioned algorithms work on the basis of a range of principles as mentioned in the following list:

- Feature-selection, statistical, or feature scoring method to choose the right characteristic to maintain in the training datasets
- Matrix factorization is another principle that breaks the parts of the features of a data point into the sets of matrices and it selects the most suitable matrix for a data point
- Manifold learning is another important methodology-based algorithm that uses different variable feature mapping techniques
- Autoencoder method is also a dimensionality reduction algorithm that is used in deep learning models of artificial intelligence projects.

Ensemble Methods

Ensemble learning methodology is a type of approach used for machine learning projects. It is the combination of multiple ML algorithms used for a particular machine learning project. This technique is used to improve the results or outcome value of an ML-model because it uses multiple algorithms to work simultaneously so that weaknesses of one algorithm are covered through the upsides of another one. Thus, you get the best results by using the ensemble method of combining multiple algorithms. The real-world use cases of ensemble ML model include digital weighted voting, averaging applications, and so on.

Transfer Learning

Transfer learning is a very useful method of transferring the learning expertise of a machine that was developed earlier for a similar type of purpose to another ML-model to learn from the previous experience learning of the past ML-project. This technique is based on the reuse of the previous artificial intelligence knowledge gained by the past machine learning systems in the new projects that also deal with the similar kinds of problems. Transfer learning uses two major approaches for carrying out the transfer of learning from past project to new one such as:

- Pre-Trained Model Approach

- Develop Model Approach.

This model is extensively adopted in the deep machine learning projects in large scales. The companies like Google, Microsoft, and others have already benefitted from this model heavily through a range of platforms of deep learning.

Natural Language Processing (NLP)

Natural language processing, precisely referred to as NLP, is a sub-domain of machine learning. This is the computational process, which enables the machines to understand, read, and derive meaning from the data input in natural language in the form of text or speech that humans communicate with. The natural language processing is a comprehensive process that incorporates the capabilities of computer science and artificial intelligence on the training datasets in the form of text and speech/voice. In other words, we can say, natural language processing is the sub-set of three areas—natural language that human beings speak, computer science, and artificial intelligence technologies as shown as an intersection of three domains in Fig. 3.1.

The main features, uses cases, working principle and other characteristics of NLP are mentioned in the following list [129, 130]:

- NLP domain is divided into three major categories as mentioned below:
  - **Natural language understanding**—NLU refers to the capabilities of the system to understand the natural language that a human speaks
  - **Natural language generation**—NLG refers to the capability of ML-models to generate language against the desired communication task
  - **Speech recognition**—The capability of NLP-powered machines to convert the speech into written text.

Fig. 3.1 Schematic diagram of NLP (self-drawn)

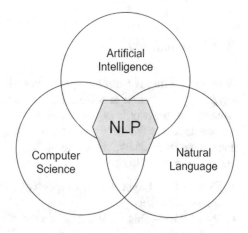

- The 3 basic categories of activities involved in NLP are speech recognition, natural language understanding, and natural language generation respectively.
- The sub-activities in a sequence to materialize the NLP field include the following:

  - The speech data is broken into small pieces known as phoneme of about 10–20 s for analysis through Hidden Markov Models (HMM)—in most cases to determine what sentences are spoken in the speech.
  - The sentences are then segmented into small parts on the basis of punctuations. This process is called segmentation.
  - Each word of the sentence is separated. This process is known as tokenization in NLU.
  - Unnecessary words that make our sentence more cohesive but do not add to the core meaning of the sentence are removed. This process is called removing stop words.
  - The word stemming to create multiple options by adding affixes to the word is done. This process is known as word stemming.
  - The next process is the "root stemming" in which all the words available in dictionary are linked to the root words such as 'are', 'is', and 'am' belong to root word 'be'. This process is known as lemmatization.
  - Parts of speech tagging is the next step in this process. In this stage, all words of a sentence are tagged with their respective parts of speech annotation.
  - Name entity tagging is the last step of this analysis process. In this process, the popup culture references are providing for different words to offer deeper understanding of the keyword to the machine learning algorithms.
  - After those preprocessing steps, the text is fed to the NLP algorithm, mostly Naïve Bayes algorithm, which creates the result of your analysis powered by the NLP applications respectively used in different platforms.

- The most common industry use cases of NLP applications include language translation tools, speech to text converters, chatbots, virtual assistants, autocorrection tools, and so on.

Word Embedding

Word embedding is a commonplace word in natural language processing. This is a type of presentation of a word that resembles to the characteristics of the representation of the words of the same meaning. Thus, word embedding means the representation of the same meaning words in the same way for NLP algorithms to understand them easily. The entire process of word embedding is governed by three major algorithms [131]:

- **Embedding Layer**—used to specify different parameters such as vector size, space, and others
- **Word2Vec**—This is Word 2 Vector algorithm developed by Google for a standalone word embedding purpose

- **Global Vector (GloVe)**—This is another major algorithm used for word embedding purpose, which acts as an extension to the Word2Vec algorithm.

What Is a Machine Learning Algorithm?

Algorithm is the most important component of artificial intelligence, especially of its sub-category named as Machine Learning (ML). An Algorithm in this case is a logical set of rules that takes raw data as an input from different sources and processes it to generate a desired output through a computer, AI program, or a neural network [92, 93].

In other words, a set of programmed instructions based on the logical solution given to an artificial intelligent system to help it for learning on its own, is known as algorithm. The definition of an AI algorithm can be expanded in future such as: a set of rules based on certain mathematical and logical calculations to solve a problem through intelligent machines like neural networks, AI programs, and other intelligent systems.

There are many different types and categories of AI algorithms, which are adopted in different AI-powered solutions. Some algorithms process the data for providing the outcome out of that data, while the others process the data to make the machines learn the patterns of input and output simultaneously. Later types of algorithms are referred to as machine learning algorithms, which are mainly powered by the training datasets. The training datasets will be covered later on in this book.

The core problem domains that are solved by using a range of artificial intelligent algorithms include regression, clustering, and classification. These categories of problems pertaining to AI can be solved through different techniques and sets of rules to generate output or results in the form of either categories, clusters, or regressions. Let us have a deeper dive into different categories and corresponding algorithms commonly used in artificial intelligence.

Common Categories of Machine Learning Algorithms

As mentioned earlier, an AI algorithm is a set of rules and techniques to solve a range of classes or categories of problems pertaining to classification, clustering, and regression as listed below. The details of associated algorithms to each of the following categories of problems will be discussed in the subsequent topics [93].

- Classification Algorithms
- Clustering Algorithms
- Regression Algorithms.

Classification Algorithms

Classification types of algorithms are those ones, which are designed to solve the problems pertaining to categorization or classes of the dependent variables in classes and then to predict the class for the given input to the AI machines. These types of algorithms use supervised machine learning type by using numerous algorithms. A few very important algorithms of this type are explained here:

Naïve Bayes

Naïve Bayes algorithm is based on the rule of probability proposed by Naïve Byes. There are numerous versions of Bayes theorem that are used in different algorithms and other domains of artificial intelligence. The Naïve Bayes theorem used for machine learning algorithms includes the following versions [93, 94]:

- Naïve Bayes Classifier Rule
- Bayes Optimal Classifier Rule.

These rules define the relationship between the prior probability function, posterior probability function, Likelihood function, and evidence function as shown in the following formula:

$$\mathbf{P(A|B) = P(B|A) * P(A)/P(B)} \tag{3.1}$$

$$\mathbf{P(B|A) = P(A|B) * P(B)/P(A)} \tag{3.2}$$

In the above probability formula, the meanings of the functions are explained in the forms of probability terminologies:

$$\mathbf{P(A|B)} = \text{Posterior probability}$$
$$\mathbf{P(A)} = \text{Prior probability}$$
$$\mathbf{P(B|A)} = \text{Likelihood}$$
$$\mathbf{P(B)} = \text{Evidence}$$

Thus, it can be expressed that Naïve AI algorithm is based on Naïve Byer Classifier Theorem, which is extensively adopted in the machine learning domain of AI. In this probabilistic approach, the input entities are fed into a possibility prior table. On the basis of that table, frequency table is formed based on the number of possible options commonly known as prior probability function. When new data are fed into the algorithm, it identifies the class in the pre-manipulated frequency table and updates the table. On the basis of frequency table, the likelihood table is created on the basis of Naïve Bayes theorem of probability.

Decision Tree

Decision tree is a step-by-step logical tree starting from the stem passing through the branches and each external node is incorporated as a test. The possible results of the test are represented through branches, which are also referred to outputs. This tree continues till the final attribute value is achieved for a particular test. That final attribute is also known as the leaf node. The schematic diagram of decision tree algorithm used in different problems pertaining to classification in customer support automated system is shown in Fig. 3.2.

The most common features and characteristics of decision-tree algorithm are mentioned in the following list:

- It is very easy and intuitive algorithm in terms of interpretation of the solution to the problems.
- This algorithm can be used for both classification- and regression-based categories of problems simultaneously.
- It is equally effective for linear as well as non-linear data.
- The construction and expansion of this algorithm is very easy and fast due to simple splitting of the nodes into the branches with the outcomes of the test attribute nodes.

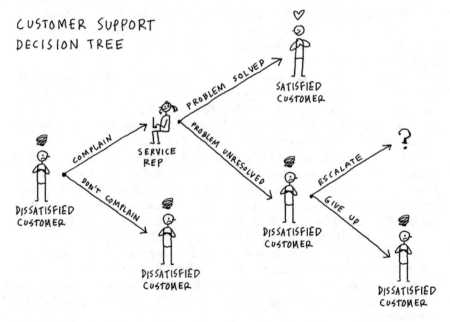

Fig. 3.2 Schematic decision tree diagram for customer support escalation matrix (Flickr)

Random Forest

Random Forest is a type of algorithm that consists of multiple decision trees acting as subset of an algorithm acting under the given (certain) conditions. The final decision is made on the basis of the outcome of the identifier trees. This algorithm is more accurate as compared to the standalone decision tree. The final decision can be made on the basis of majority results of the subsets incorporated in a random forest fashion.

Support Vector Machines

Support Vector Machines, precisely referred to as SVM, is a type of algorithm extensively used in supervised machine learning systems. It is an algorithm that helps solve the problems related to the identification of the entities or items. The use of SVM is very high in natural language processing (NLP) for text recognition and similar kinds of applications. It is very effective for a limited data, in thousands of entries for example, due to its faster speed and better performance.

The most basic principle of operation of support vector machine is based on finding the maximum margin separation between the two categories in a multi-dimensional plane. It draws a hyperplane which is either a two-dimensional or a three-dimensional separation line between different classes of entities that is used to separate them on the basis of a formula that determines the maximum distance or margin between the entities of different classes.

The hyperplane is drawn in such a way that it determines the largest possible margin between items in both categories. This hyperplane divides two classes of the items or entities in this plane. The distance between the nearest object in class A should be the possible maximum from the hyperplane line and similarly, the distance between the hyperplane line and the class B objects should also be maximum.

There is only one such possibility, which is determined through mathematical formula. The maximum distance between the nearest object in class A and the hyperplane line is known as margin A. Again, the maximum distance between the nearest object in class B to the hyperplane line is referred to as the margin B. Both margin A and margin B should be maximum in any other possible calculation in this system. It is considered as the best hyperplane line as shown in Fig. 3.3 [95].

In the above-figure, two-dimensional data and separation of the data are considered in the basic form, which is also known as linearly separable binary sets. The half-circles are the object class A and the full-circles are the object class B. The support vector machine algorithm uses the following formula to compute the linear hyperplane:

$$\mathbf{W}.\mathbf{x} + \mathbf{b} = \mathbf{0} \tag{3.3}$$

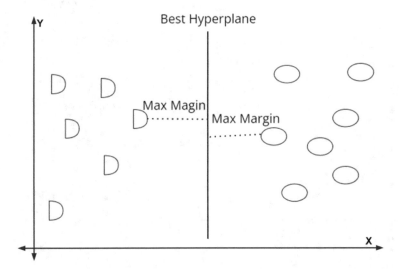

Fig. 3.3 Schematic diagram of support vector machine algorithm

where:

W is the vector perpendicular/normal to hyperplane
x feature vector
b offset value based on definition rule.

This is very important to note that the above-mentioned example is just an explanation for linearly separable binary sets. Normally, data in the real-world is not that simple. It is spread in three dimensions and may appear in haphazard order. In such conditions, the data is analyzed in three dimensions and then a non-linear hyperplane is found based on a more complex formula.

For the classification of non-linear and other forms of complex data, the support vector machine algorithm uses the following additional functions and mathematical tricks:

- The Kernel Trick
- Kernel Function.

K Nearest Neighbors

K Nearest Neighbor, precisely referred to as KNN algorithm, is another very important one in the field of artificial intelligence, especially machine learning domain. It is used for both classification and regression related solutions. But the use of KNN algorithm in classification problem solutions is more pervasive than in the regression-based solutions. This algorithm is considered as one of the simplest algorithms used

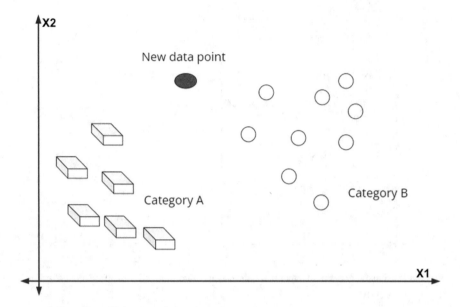

Fig. 3.4   Schematic diagram of K nearest neighbor algorithm

in machine learning training datasets. It is also known as non-parametric algorithm. The meaning of non-parametric algorithm refers to the meaning that it does not use the concept of assumption for the underlying data points. The detailed working principle of KNN algorithm is shown in Fig. 3.4 with a schematic diagram [96].

The basic function of its operations is to find the similarities of the new data point (input) with the characteristics of existing categories to decide about the category of that new data. For that purpose, it chooses a few numbers of neighbors referred to as "K". It is an arbitrary value. Then, it measures or calculates the Euclidean distance of those nearest neighbors equal to K numbers on both categories of information. Now, it counts the number of data points in each category. The relatively larger number of data points that fall in the pre-selected number (K) range is considered as the category of the new data point in this machine learning algorithm of data classification.

The most common features and characteristics of K nearest neighbor machine learning algorithm is mentioned below:

- It is based on supervised machine learning technique.
- This algorithm stores the data continuously and compares with the existing data for classification of the next data points.
- Finds similarities in the new data input with the existing categories of data to categorize the new data in the system.
- This algorithm is also popular by the name "*Lazy learner algorithm*", which means that it does not learn from the datasets at the time of input or immediately. It stores the data point in the database and performs the identification process when the need of classification of the data points arises.

- As the database increases the performance, speed of this algorithm gets slower.
- The linear distance between two points is known as Euclidean distance, which is measured through the following formula in a two-dimensional paradigm:

Euclidean Distance between point A(X1, Y1) and Point B(X2, Y2)

$$= \sqrt{(x2 - x1)^2 + (y2 - y1)^2}$$

- The K value is selected by the users of the algorithm. It should be selected in such a way that the outcome of the selected data has minimum error ratio or noise in the output. The most commonly used K-value is five (5). Using less than this number is prone to errors or noise in the data and higher rates may affect the speed of the machine. The higher value is always desirable for getting more accurate data for classification.

Clustering Algorithms

Clustering algorithms are the mathematical techniques used to group the data points into different natural clusters or groups. There main purpose of clustering algorithms is to materialize the concept of clustering in artificial intelligence, especially in machine learning systems. The clustering is also known as cluster analysis in AI domain. It is a type of un-supervised type of machine learning in which computer machines or intelligent systems are not trained on the basis of tagged datasets rather they learn themselves from the given data on the basis of similarities of a range of characteristics and attributes of the data fed into the intelligent systems to create corresponding natural groups or categories [97].

The basic principle to operate in all types of algorithms in clustering is to analyze the data in terms of a range of characteristics of the data such as color, behavior, size, interests, activities, and many other attributes and form a number of natural groups of the data in line of the similarity of those characteristics in maximum numbers. This technique can classify a data point into one or more than one group based on certain similarities in their respective behaviors. Clustering algorithms perform a range of technical tasks used in the cluster analysis. A few of the most important tasks that are commonly performed by the clustering algorithms include:

- Statistical data analysis technique
- Image segmentation technique
- Market segmentation technique
- Social network analysis technique
- Anomaly detection technique.

The examples of the application of clustering include the recommendations that appear on your page while visiting certain e-Commerce websites like Amazon, Walmart, and others. They use the clustering techniques to get the most suitable recommendations for you on the basis of your interest, age-group, locality, profession, and so on. The above-mentioned tasks are performed through a certain technique or some technical methods that are implemented through algorithms in cluster analysis. A few of those clustering methods include:

- Partitioning-based cluster analysis
- Density-based clustering
- Distribution model-based clustering analysis
- Hierarchical clustering analysis
- Fuzzy clustering.

Those methods are materialized through different algorithms. The most common clustering algorithms used in the modern field of machine learning/artificial intelligence are described below [97, 98].

K-Means Clustering

K-means clustering algorithm is one of the most popular algorithms used in the modern machine learning processes. In this algorithm, the data is divided into the pre-defined groups based on the equal variance. The number of groups or clusters is provided arbitrarily. This is also very simple and faster type of algorithm for clustering the data points. This algorithm uses the partitioning-based clustering analysis method for categorization of the data points.

Expectation Maximization (EM) Algorithm

This algorithm is designed to maximize the expected attributes or characteristics of a cluster of data points. There are two major categories of attributes of a data point—observable and latent variables. The observable variables can be detected through different techniques but the latent variables of a data point are those hidden attributes that need to be achieved through a robust algorithm like expectation–maximization (EM) algorithm. This algorithm is based on two major types of repetitive steps as mentioned below [99]:

- Expectation Step (E-Step)
- Maximization Step (M-Step).

Both of the above-mentioned steps are repeated until the convergence of the attributes is achieved. The first step assesses (guesses) the missing data values while the second step updates the data parameters of the clustering achieved in the second step.

Agglomerative Hierarchical Clustering

This algorithm is designed to perform hierarchical clustering from bottom-up. In this algorithm, all the data points are initially considered as a separate cluster and then merged into each other on the basis of their attributes and characteristics. The merger is done through a tree-structure, which provides a better relationship based on the hierarchical order.

Fuzzy C-Means Algorithm

Fuzzy C-Means algorithm, sometimes referred to as Fuzzy K-Means algorithm is designed to perform fuzzy clustering analysis. This algorithm allows multiple data points to be parts of a range of groups based on their respective membership coefficient. In this algorithm, the membership coefficient is determined for a data point, which is member of multiple groups or clusters. The membership coefficient defines the characteristics of that particular data point, which can be member of multiple groups simultaneously. The membership coefficient is the distance value from the center of the cluster or group of data to the center of the data point from each cluster of which the data-point is member in the data-ecosystem.

Regression Algorithms

Regression algorithms are those that estimate or predict the dependent variable(s), which is also referred to as target variable or output, in relationship with many independent variables that can impact the value of the target variable. This technique is used for training the machines through supervised machine learning scheme to enable it to predict the desired outcome by learning the relationship between a range of variables in different situations through regression of the parameters to generate a continuous range of parameters in numbers.

The most common usages of the major types of algorithms based on regression technique include the following:

- Forecasting market trends
- Projecting stock markets and other businesses
- Price, salaries, costs, consumptions, and many similar kinds of values in given conditions.

There are numerous types of algorithms that use the regression mathematical technique to generate the target value. A few very important ones are covered in the following sub-topics.

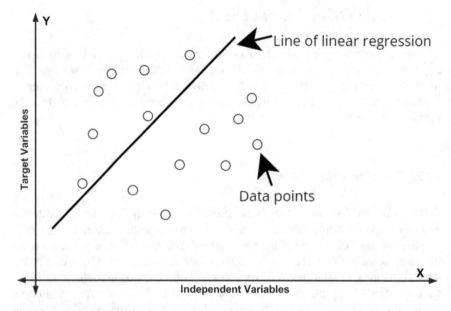

Fig. 3.5 Schematic diagram of linear regression algorithm

Linear Regression

As the name implies, the linear regression predicts the trend of target value in a straight line drawn with respect to the impact of an independent variable (data point). This type of algorithm generates the most fit line to depict the target value under the given range of independent variables. It is much simpler in deployment as well as in data processing.

The schematic explanation of linear regression where the independent variable is projected on the X-axis dimension and the dependent or target variable is project on the Y-axis is shown in Fig. 3.5.

In the above-diagram, the straight line in bold provides the best possible value of the target value in response to the given independent variables. The small circles show the values or positions of the data points fed into the algorithm for prediction.

Multiple Linear Regression

In multilinear regression, one target value is determined by following the same technique as used in the simple linear regression with more than one predictor variables or independent variables. The simple linear method predicts only one dependent variable with one predictor variable or independent variable. Thus, multiple factors or linear lines are projected in multiple linear regression model. All other parameters, techniques and factors are almost similar to the simple linear regression algorithm.

Multivariate Regression

Multivariate regression is another type of algorithm used to predict multiple outcomes or target variables in terms of multiple predictors. This model is more effective in the situations when multiple independent variables are influenced by the dependent variable(s). There are certain shortfalls in simple and multiple regression algorithms— and, this method covers those downsides. It uses multiple independent variables and their respective impact on the dependent variable or output variable. Thus, we can say that multivariate is an extension of multiple linear regression with additional impact of multiple independent factors on the dependent variable [100]. Thus, you get an output line with impact of those factors making the line move in discrete directions.

Logistic Regression

Logistic regression is another very important type of regression-based algorithms. It is extensively used in statistical analysis for determining logical outcomes in the form of two states. The output of this model of analysis or algorithm is generated in binary formats, either 0 or 1. Thus, it generates only two target variables in relationship with different dependent variables to determine the state of that particular outcome.

Lasso Regression

The Least Absolute Selection Shrinkage Operator, precisely referred to as LASSO, is a type of regression-based algorithm, which puts the constraint or shrinkage on the attributes or parameters to reduce them to zero or near zero. This model is designed to identify and apply the shrinkage or constraints on those variables whose coefficient tends towards the zero or zero. Such variables are struck off in the regression calculation. In other words, this model is used to eliminate the variables from the data analysis that are of less importance with zero coefficient.

Other Regression algorithms

A few most important models based on regression technique have been explained in the above topics. There are many other algorithms based on regression calculations that are used in different types of applications. A few of them include:

- Poison regression algorithm
- Ridge regression algorithm
- Polynomial regression algorithm
- Bayesian linear regression algorithm
- Quantile regression algorithm
- Elastic net regression algorithm
- Principal component regression.

There are also a few other algorithms that are used in the training datasets for more than one domain of analysis. The following models are used in both regression and classification analysis training sets in the modern artificial intelligence and machine learning field:

- Decision tree regression algorithm
- Random forest regression algorithm
- Support vector regression algorithm.

What Is AI Training Data?

Training data is a fundamental component of machine learning (ML), which is a major field in artificial intelligence (AI). Thus, we can say, training data plays a pivotal role in realizing the modern concepts of artificial intelligence or achieving the human-like thinking in computer or computing machines. Let us define here what the training data is.

Any kind of data in the form of text, audio, video, image, or senses like thrust, temperature, light, touch, smell, and others that can be used for training the artificial intelligence models integrated with the machines so that those machine-powered prediction models can learn, understand, and decide about the future actions is called the training data [101–103].

The training data is fed into the computer machine-based models through two different ways such as:

- Labeled and annotated input
- Unlabeled raw data.

The labeled data is fed into the supervised machine learning models. The type of input data that is labeled with the detailed information of the parts and components of that dataset is called labeled data set. The supervised learning requires training through labeled and annotated data. At the other hand, the raw data without any labeling or annotation is fed into the machine learning models based on unsupervised machine learning.

Both supervised and unsupervised machine learning that use a range of training data for learning about the real-world environments include the following:

- Text Training Data
- Audio Training Data
- Video Training Data
- Image Training Data
- Sensory Training Data.

As mentioned earlier, the supervised machine learning-based models use labeled/annotated datasets, which are commonly referred to as training datasets or learning datasets. The learning datasets can further be divided into three major categories with respect to their purposes:

- **Learning datasets**—These datasets are used for initial training of machines to understand how to apply different machine learning technologies such as neural networks and others to analyze and generate sophisticated results.
- **Validation datasets**—Those datasets that are used for validating the previous training and level of learning capabilities of the ML-based model to ensure that the training is serving the desired objectives of the project.
- **Testing datasets**—Like validation datasets, which are also a part of testing training, these are used to check if the training is achieving the desired goals of the models.

Types of Training Data

The major types of training datasets used in both supervised and unsupervised machine learning are explained below [101–103].

Text Training Data

Text is one of the largest and very effective sources of information in the world. In most of the cases, the text data is highly unstructured that the machines cannot understand them easily. Another major problem related to the building of text training sets includes a wide range of languages, different formats of text such as digital, printed, and scanned texts. Hence, providing detailed classifications of the grammar, meaning, sentiments, and other parameters of the text is known as text training dataset formation for supervised machine learning.

Audio Training Data

Another training data used for machine learning projects is audio. In this type of data, files of voice, music, and other types of sounds available in the real-world are consumed in the machine learning models through different types of formats of files. Those different formats and types of audio data are fed into the ML-powered projects.

Video Training Data

Video has become a very crucial form of data in the modern artificial intelligence and machine learning projects. With the advent of modern video technologies and high-speed connectivity, the use of video for a range of modern AI/ML project has become very pivotal. The modern driverless cars are highly trained through video and image training data for effective learning, analysis of the scenarios, and making suitable decision with greater level of accuracy.

Image Training Data

Incorporating details of images into the ML-powered supervised machine learning with proper details is called image training data. The unsupervised ML uses images without labels and annotation for deep learning purpose. A wide range of formats and types of images are used for image-based training datasets for AI projects.

Sensory Training Data

Sensory training data comes in the forms of different units, files, and formats. This data is transmitted through different sensors to the ML models. The major types of sensory training data sets in different forms and formats include thrust, push, feel, light, smell, smoke, moisture, and many others.

What Is AI Training Dataset?

A range of types of training data have been discussed in the above topics. Those different types of data sets are in unstructured formats, which is difficult for machines to learn and understand. Thus, those data types are labeled, classified, tagged, and annotated in different formats, attributes, tools, and classes in such a way that the machine learning algorithms can easily understand them and keep in their artificial mind (database) for making future decisions regarding the similar kinds of situations or conditions in the real-world problems. Such structured files of data are known as artificial intelligence or machine learning training datasets.

Major Processes Used in Building Training Datasets for AI Training

The formation of artificial intelligence or machine learning-based training datasets involves a range of processes that are performed by human resources with the help of numerous tools and platforms. The most common processes for building professional-level training datasets for AI/ML projects include:

- Data Collection
- Data Cleaning
- Data Classification
- Data Categorization
- Data Annotation and Labeling.

The details of all those processes used for creating training datasets are mentioned separately in the following sub-topics.

Data Collection

This is the very first process for building training datasets. The collection of data refers to gathering of required data to be used for training purpose of the ML-projects from a range of sources such as online content, creating real-world videos, taking from partners, accessing local databases, and many others. A few important things that should be taken care of while collecting data are quality of data, relevance, and balance of data types [104].

Data Cleaning

The second process of forming specialized datasets for machine learning and deep learning project training is the cleaning of the data. Mostly, the unstructured data collected from different sources contains a huge volume of noise, irrelevant items, duplication, absence of required types of data, biased data, and other impurities, which are supposed to be cleaned to make data more useful for creating datasets.

The entire process to make collected data highly useable to achieve the desired training objectives is known as data cleaning [104].

Data Classification

It is a process to develop training datasets for machine learning in which each given set of data is classified into categories. This can be done through different models or algorithms of classification such as binary classification, multi-class and multi-label classification, and many others.

Data Categorization

Data categorization is a process to build training dataset for supervised learning in which the data is classified in terms of groups, types, levels, and other larger attributes so that machines can understand the input training datasets and can consume them for machine learning and deep learning purposes. Content categorization, text categorization, and item classification are a few very important examples.

Data Annotation and Labeling

Data annotation and data labeling are two separate terms but extensively used inter- changeably due to very bleak separation line between those two concepts. The data annotation refers to the techniques used for labeling the data, especially images and videos, for creating a tagged dataset for machine learning training. On the other hand, the data labeling is the concept of tagging any kinds of data with a range of attributes of that item. Thus, data annotation is a part of data labeling [105].

What are the Major Categories of Data Annotation?

As mentioned earlier, data annotation is a subcategory of data labeling for building a training dataset for AI-systems to learn through supervised learning type of artificial intelligence. With the general understanding, there are five major categories of data, which will be further explained with different types of techniques used (in those categories):

- Image Data Annotation
- Text Data Annotation

- Audio Data Annotation
- Video Data Annotation
- Sensory data annotation.

The sensory data is normally fed through different formats either in the form of image, video, or in other units pertaining to the related sensor. Thus, we may not consider the last one as a distinctive type because it is not annotated in any type of data other than the four (other) major categories for supervised learning technique (i.e., sensory data would eventually fall under any of the other four types).

Image Data Annotation

Identifying, labeling, and categorizing the image and its items is called image data annotation process. It is done through numerous tools, techniques, and through text content. The image data annotation is further divided into different techniques as mentioned below.

Bounding Box Annotation

In this type of image annotation, a bounding box is used in an annotation tool to surround or bound the entire body of an item in an image. This bounding box is a two-dimensional square used for circling the boundaries of an item. This is one of the most fundamental techniques of annotation, which is extensively used in both image and video annotations.

In bounding box annotation, an annotator should ensure the following key things while conducting annotation activities.

- Minimum number of overlaps
- Minimum fluctuation in sizes of boxes
- Reduced size of box overlaps.

A simple example of bounding box image annotation is shown in Fig. 3.6.

3D Cuboids Annotation

In a 3D cuboid image annotation, a 3-dimensional cube is used to scale all three dimensions of an image to produce more accurate results. This is more useful when the accuracy of item size and distance between the items matters a lot.

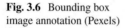
Fig. 3.6 Bounding box
image annotation (Pexels)

Polygon Annotation

All objects are not regular in shape that can be mapped under squares or cuboids. To mark all edges and other features, polygons are used to generate more accuracy in the shape and size. Using polygon technique is called polygon annotation.

Lines and Splines

Lines and splines are used to determine the boundaries of different jurisdictions, activities, or items to differentiate from one part of an image from the other part of the image. This annotation is extensively used in both video and image annotation for roads, streets, traffic control systems, and others.

Semantic Segmentation

In this technique, each and every part and point of an image is assigned a label for better understanding of the landscape depicted in the image. Multiple similar types of points or items are treated as the same in this technique and marked with the same label or color.

Text Data Annotation

Identifying and assigning the labels to the text in a digital file is called as text data annotation. It is used for supervised machine learning in artificial intelligence. A digital file can be a scanned copy, text file in different formats, and others. There are five major text annotation classes that use different sub-types of text annotations as mentioned below.

Entity Annotation

In this text data annotation technique, the names of place, person, value, time, distance, and other such categories are identified and assigned. A few sub-functions include:

- Named Entity Recognition (NER)
- Key Phrase Tagging
- Parts of Speech (POS) Tagging.

Entity Linking

As the name implies, in this technique, the entities in the text are identified and linked with the category of that entity through a URL link to make it more understandable. The entity linking becomes a very effective source for the machines to learn about that concept deeply. A few major types of entity linking technique are listed below:

- End to End Entity Linking
- Entity Disambiguation.

Sentiment Annotation

The labeling of text words and signs with associated sentiments of the writer are marked and labeled in this type of text annotation technique. With the help of sentiment annotation, the data annotators train the machines to understand the emotions behind the writings or words.

Text Classification

Determining and tagging an entire document with a single word, phrase, or a short sentence in terms of its category, class, tone, sentiment, or any other tags is known as text classification. This applies to the whole text as a single item. A few sub-topics of text classification include:

- Document classification
- Product categorization
- Sentiment annotation.

Audio Data Annotation

Audio data annotation is a process of labeling different types of audios with predefined attributes such as names, quality, noises, sound types, frequencies, emotions, and others. An audio file may include speech and other sounds in our real-world such as machine noise, music, bird-sounds, natural voices, and so on. Different types of audio annotations used in this category are mentioned in the following sub-topics [106].

Sound Labeling

Adding additional information to the sounds or speech in the shape of metadata and other tags and notes is called sound labeling. Adding additional labels enable machine learning projects to understand the reasons, objectives, need, and many other factors associated with that particular format or nature of sound.

Event Tracking

Event tracking is a technique to highlight certain events of sounds in an overlapped sound environment so that machines can understand the sound events that occur in a complex sound consisting of multiple sources and frequencies in one speech or file of other sounds.

Speech to Text Transcription

The conversion of speech into text is called the text transcription. This is also one major type of training dataset to enable machine learning projects or systems to learn about the speech through a comprehensive text. This method plays a very vital role in modern projects such as chatbots, virtual assistants and other applications powered by natural language processing technology (NLP) in the field of machine learning.

Audio Classification

Audio classification is a process to tag the sounds for distinguishing between the speech instructions and sound instructions for a machine learning project. A range of commands are fed into the machine to distinguish between the voice and speech commands.

Multi-labeling

Multi-labeling of sound or speech file is the process of tagging multiple sources of data with different predefined categories. This type of labeling of audio data files is further divided into two main categories such as [107]:

- One-stage multi-labeling
- Two-stage multi-labeling.

Video Data Annotation

Video annotation is a process in which the video clips are tagged or labeled frame-by-frame to generate a training dataset for machine learning and deep learning projects. This process is very useful in modern computer vision (CV) application, which is one of the most popular technologies in modern artificial intelligence-based projects.

It is almost similar to the image annotation with a major difference that the video annotation is accomplished on a clip that consists of a large number of frames. In other words, those frames act like an image. Labeling of the video is done on the basis of predefined item categories that appear in a video clip. The main objective of video data annotation is to build training datasets for ML/DL projects with main focus on achieving the following objectives:

- Detecting objects appearing in video by machines
- Localization of the moving objects appearing in the clip
- Tracking the position and status of the object continuously

- Tracking and monitoring the activities of the objects in the video.

There are numerous types of video data annotation used in the modern world for annotating the video data or clips. All major types of data annotations used in the image annotation process are also used in the video data annotation such as:

- 2D and 3D bounding box annotation
- Semantic segmentation annotation
- Polyline and polygon annotation.

Other than the above-mentioned types of annotation techniques, a few additional types of video annotation are mentioned below [108, 109].

Key Points Annotation/Landmarks

Key points annotation, also referred to as landmark annotation, is a type of video annotation to mark smaller features of an object, or smaller things in a video and connect them to form postures, sentiments, movements, and other attributes. This type of annotation plays a very vital role in precision application based on computer vision such as facial recognition, sporting analytics, and other similar kinds of applications.

Object Localization

Object localization is a type of video annotation in which one or more than one items of similar categories are identified and tagged with that particular class and surrounded with a particular marking sign such as 2D bounding box around those objects. This function helps computers and intelligent systems identify and track the objects in an effort to better understand the real-world environment in motion.

Object Tracking

Object tracking is an application of the video annotation that tracks the moving objects based on certain pre-trained video annotation techniques. It is extensively used in applications that are designed for identifying and locating certain objects in a video application through camera. The use of object tracking function is extensively noticed in the security and surveillance applications in numerous parts of the world for maintaining peace and security in an area. This function is also used in the management of traffic signals, medical imaging, and many other applications.

Gradient Boosting

Gradient boosting is one of the powerful machine learning algorithms used in numerous predictive models in two major categories of problem solutions named as classification and regression. It is used in both of them equally because it relates to the decision tree algorithm, which is used in both of the categories of problem solutions in machine learning. It is especially used for the supervised learning of the machine learning models and also known as greedy algorithm because of its capability to overfit a training dataset faster. The gradient boosting machine learning algorithm was developed by Jerome H. Friedman on the basis of prior work of Leo Breiman.

Gradient boosting algorithm is used to form ensemble machine learning model by combining more than one weak learner model to improve the prediction performance, especially with the decision tree techniques. The weak learners provide boosting or improvement impact on the weak predictive models due to this combination through a mathematical formula. In this method, a weak decision tree is ensembled to make it gradient-boosted tree, which is done in step-by-step fashion with a predefined order [114].

The gradient boosting algorithm incorporates three major elements which are listed below:

- **Loss function**—An arbitrary differentiable loss function can be used in this algorithm that is supposed to be optimized
- **Weak learner**—A weak learner is made to conduct predictions in this model
- **Additive model**—Used to add weak learners to reduce the loss function.

The improvements in the basic gradient boosting algorithm achieved in general applications are mentioned in the following list:

- Tree constraints
- Shrinkage or weighting updates
- Random sampling or stochastic gradient boosting
- Penalized learning or gradient boosting.

Thus, the above-mentioned different types of gradient boosting or enhancements lead to sufficient improvement in the prediction precision, especially in decision trees, which are considered as the weak learners in ML algorithms.

Top Uses of Machine Learning in Today's World

According to the Research and Markets forecast, the global market size of machine learning is expected to cross $27.7 billion by 2027 from just $2.8 billion in 2020 with a gigantic growth of over 38.4% over the projected period [134]. This explosive growth is not driven by a few industries but across all the domains of industries and businesses in the world. A few very important business domains where the impact of machine learning is noticed the most are explained in the following sub-topics [132, 133].

Big Data

Big data has emerged as one of the most powerful technologies for managing the enormous heaps of data generated by the modern systems in all domains. The best management of big data has become possible due the power of machine learning and artificial intelligence. Big data is characterized by 3 V: Bigger volumes of data, faster velocity of generation/processing of data, and a wide range of variety of data. The entire big data applications are powered by the artificial intelligence and machine learning to dig out the hidden values of the heaps of data in numerous ways.

The scale of data produced nowadays can be just a big liability for the companies to store, maintain, and dispose of with the power of modern artificial intelligence and machine learning technologies. The big data would have been just the heaps of garbage, if the ML/AI had offered desirable capabilities of wide range of benefits to store, manage, process, analyze, and skim out the valuable information related to patterns and behaviors of the users and businesses.

Data Analytics

Modern data analytics is extensively governed by the power of modern ML-powered applications. The data analytics field has become so dynamic and fast that it can produce results in the real-time environment that are helpful for a range of businesses. The most common benefits achieved through data analytics powered by machine learning include:

- Better insight into the behavior of different segments of markets and user trends
- Accurate forecast about the future demand and supply of the concerned services and products
- Accurately targeted marketing and advertisement for effective reach out
- Improved business operations through data-driven policies and strategies
- Efficient use of available resources to make most of them
- Achieving the best user experience and customer satisfaction.

Cybersecurity

Cybersecurity is a very important domain where a wide range of machine learning algorithms and systems are used for improving the cybersecurity issues and help the cybersecurity appliances and applications to produce more robust security to the networks, applications, and devices in a range of computing and networking environments. The safeguarding of the business communication through emails, online collaboration platforms, and other mediums is secured by the use of different applications powered by the machine learning technology. The most common uses of machine learning in cybersecurity include:

- **Spam filters**—Useful in maintaining secure communication through emails and other web-based communication sources
- **Vulnerability assessment tools**—Numerous types of vulnerability assessment tools are playing vital role in detecting and preempting any possible weaknesses in cybersecurity of the existing systems or applications
- **Automated security response algorithms**—There are many machine-learning-enabled and automated algorithms that respond in certain security situations to avert emerging threats.

Digital Marketing

Modern digital market is heavily dependent upon the business intelligence about the products, businesses, market trends, customer segments, user behaviors, customer feedback, and many other such parameters that measure the success of modern marketing—subsequently, the business in today's fiercely competitive and knowledge-driven environment. The most common domains of digital marketing impacted by the power of machine learning include:

- Search engine optimization (SEO)
- Targeted email marketing
- Right content marketing
- Social media marketing
- Paid advertisement.

Business Intelligence

Business intelligence is the most fundamental component of all types of modern businesses because every business has become so influenced by information or knowledge. Without knowledge of what, where, how, when, and other aspects of actions, no business can survive. This knowledge is known as the business intelligence. In the old days, the business intelligence was powered by just the human acumen, experience, and feelings. But the existing businesses are fully dependent on the business intelligence skimmed from the data generated from different sources with the help of numerous ML/AI-powered applications. Those applications help gather business intelligence from different sources and processes such as [135]:

- Customer support and feedback
- Customer behavior and interest analysis
- Surfing patterns and shopping habits
- Searching for interests, products, and other stuff
- Social media communication and activities
- Data quality checks and analysis improvements
- Cybersecurity threat intelligence and monitoring
- Automated operations and response systems.

Process Automation

Whether it is industrial process or other business process automations, the impact of machine learning and artificial intelligence is pervasive in all industries and sectors of businesses. In other words, it will not be incorrect that modern automation is fully influenced by machine learning and artificial intelligence. The most common applications where ML and AI play prominent role include:

- Manufacturing process automation through robotics
- Home automation through Internet of Things (IoT) and other sensor-based systems
- Access control and physical security systems
- Customer support and feedback systems
- Virtual assistance and remote helpdesk systems
- Product defect detection and rejection systems
- Facial recognition and voice recognition systems
- Cybersecurity threat intelligence and monitoring systems
- Automated operations and maintenance systems
- Agriculture farming automations and telemetry systems
- Automated communication response
- Driverless vehicles and Unmanned Aerial Vehicles (UAVs).

Automobiles

Automobiles have used the power of machine learning very fast and impressively during the past few years, especially in the development of driverless cars, which are going to hit the global automobile grounds extensively very soon. The entire automated car system is powered by artificial intelligence and machine learning, which enables the computer systems to learn from the real-world surroundings and incorporate into its operations (something) like a human does.

e-Commerce

e-Commerce started a few decades back just like an online platform to trade, but it grew matured very fast and within a few decades, it has become a top player in all types of modern and traditional businesses. The total volume of e-Commerce has crossed trillions of dollars by now. It has used the power of machine learning in numerous domains to expand its size and volume; a few of those domains include the following:

- Getting deeper insight into the behavior and interests of online customers
- Developing traction for online shopping like fun and enhancing customer experience
- Deployment of machine learning in a range of industries such as retail, marketing, banking, entertainment, process automation, and so on
- Efficient use of online social and community platforms and effective engagement with the prospective online shopping clients.

Impact of Machine Learning on Cybersecurity

As far as the impact of machine learning technology on cybersecurity field is concerned, it is influencing both positively as well as negatively. The power of machine learning leaves a complementary impact on the cybersecurity by improving efficiency and effectiveness of numerous cybersecurity processes through large-scale automation. On the other side, this capability of machine learning is for both the bad and good guys. Hence, bad guys can exploit machine learning to inflict damages to the cybersecurity of our systems. Let us explore both the positive and negative impacts of machine learning on the domain of cybersecurity separately [136, 137].

Positive Impact

Machine learning enables the security systems, devices, and applications to automatically analyze, detect, and respond to the emerging threats to leave numerous positive effects on modern cybersecurity such as [136, 137]:

- It makes cybersecurity systems more effective, reliable, and efficient with a very little investment to achieve the robust security with the help of numerous analytic, detective, scanning, monitoring, and reporting capabilities of ML applications.
- Enhances the threat intelligence capabilities by analyzing numerous types of data such as threat patterns, threat sources, threat frequencies, objectives and many others so that there is lesser chance of false positives.
- Faster detection of the emerging threats improves effectiveness of cybersecurity systems to act timely well before the damage goes beyond control.

- Deeper monitoring and analysis of weak points, device vulnerabilities, and reporting enable the security personnel as well as the automated systems to get the information about the emerging threats on time and take suitable actions or preemptive measures on the basis of that intelligence gathered through ML-powered cybersecurity tools.
- Enables the cybersecurity systems, devices, and applications to track the emerging threats in the real-time environment and act accordingly to counter those developments in different spheres of cybersecurity ecosystem.
- Improves the speed of scanning, analyzing, and hunting down the cybersecurity threats without wasting a lot of time on hunting the threats manually.
- User and Event Behavioral Analytics (UEBA) capabilities of machine learning algorithms enables the security personnel to define the right security policies based on proven baselines of normal users, servers, network elements, accounts, and other components that are the parts of the modern cybersecurity systems.
- Helps analyze the traffic patterns to alter any emerging security threat before it can hit the target and damage the digital resources by breaching the security systems.
- Monitoring and operations of all major processes to detect any kinds of anomaly that is suspicious under the predefined policy helps security personnel of a company monitor all types of discrepancies in the light of cybersecurity policy.

Negative Impact

The most common negative impact of machine learning on the cybersecurity that is considered by the security experts is the malicious use of power of machine learning to breach cybersecurity fences and inflict damages on the digital systems. This is because of the fact that hackers and other malicious users are highly skilled and they adopt extensively sophisticated technologies to unleash the cyberattacks on the targets that can lead to dangerous consequences. Other than this negative aspect of machine learning, the other downsides are of commercial importance such as:

- Use of costly resources
- Training of personnel for datasets and other activities
- Updating and management of modern machine learning tools
- Managing data integrity, privacy, and other sides while processing huge data under machine learning cybersecurity environments.

Sample Questions and Answers

Q1. What are the three most common machine learning functions? Describe each of them.

A1. The most common features of a machine learning activity of function can be divided into the following three categories:

- **Descriptive function**—This function of machine learning system uses the data to learn and explain what has happened in the given situation through a descriptive capability. This feature helps figure out the details of the activity or incident.
- **Perceptive function**—This capability of machine learning enables the users to get suggestions through the analysis of the data from the machine learning systems about what action is to take in the given situation of the data.
- **Predictive function**—This feature enables the intelligent systems to project about the future situation. What will happen under the given conditions based on the raw data? Intelligent systems can help through predictive capability.

Q2. What is unsupervised machine learning?

A2. In unsupervised machine learning, the unlabeled or untagged datasets are used for training the machine learning algorithms to conduct cluster analysis on those datasets. The algorithms used in unsupervised learning discover range of features and patterns without any manual human intervention.

Q3. Name the three most common categories of machine learning algorithms.

A3. Three most common categories of machine learning algorithms are:

- Classification Algorithms
- Clustering Algorithms
- Regression Algorithms.

Q4. What is object localization? Why is this used?

A4. Object localization is a type of video annotation in which one or more than one items of similar categories are identified and tagged with that particular class and surrounded with a particular marking sign such as 2D bounding box around those objects. This function helps computers and intelligent systems identify and track the objects in an effort to better understand the real-world environment in motion.

Q5. What are the most common uses of machine learning in cybersecurity? Give some examples.

A5. The most common uses of machine learning in cybersecurity include:

- **Spam filters**—Useful in maintaining secure communication through emails and other web-based communication sources

- **Vulnerability assessment tools**—Numerous types of vulnerability assessment tools are playing vital role in detecting and preempting any possible weaknesses in cybersecurity of the existing systems or applications
- **Automated security response algorithms**—There are many machine-learning-enabled and automated algorithms that respond in certain security situations to avert emerging threats.

Test Questions

1. How does Business Intelligence work?
2. How does digital marketing work?
3. How are data annotations classified?
4. AI Training Data: What is it?
5. How does a Machine Learning Algorithm Work?
6. How do Support Vector Machines work?
7. How do classification algorithms work?
8. How does Deep Machine Learning Work?
9. Do you think a machine can learn like a human being learns? Why or why not?
10. Why is Machine Learning important in the modern world?

Chapter 4
Blockchain Technology

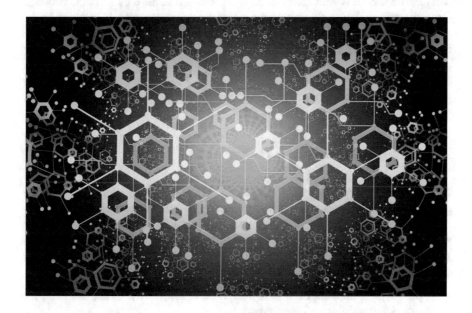

Introduction to Blockchain Technology

Blockchain technology is a digital peer-to-peer network powered by the innovative, secure, reliable software platform for the management and processing of customers' transactions without any centralized control and management mechanism. It is fully decentralized and peer-to-peer network, which is used to manage both tangible and intangible assets through digital transactions. In this technology, the software operating at the core of the platform is fully distributed, which means that there is no centralized node like servers to control and manage. This network is managed by the automated network consisting of numerous nodes located at diverse locations, which process and verify each transaction that takes place on the network to authenticate its legitimacy and correctness. This technology consists of the following components [138–140]:

- A network of peer-to-peer nodes located at diverse locations
- A distributed data management software
- Immutable records of tangible and intangible assets
- Smart contract algorithms/protocols.

Blockchain technology is so reliable and secure that no transaction of online activities can be changed or tempered to alter the correctness of the transaction in the database because it is verified by all nodes spread across the regions in the network. Once the transaction is checked and verified for its correctness, it is stored in the form of block of data, which is also encrypted into hashes. Thus, it is also referred to as the distributed ledger of immutable transactions for management of tangible as well as intangible assets. This technology was first-time developed and used for the cryptocurrency "Bitcoin", which is a decentralized system of digital currency without any intervention of any third-party or centralized body.

Top Features of Blockchain Technology

Blockchain technology is getting increasingly popular in all sectors, industries, and governmental domains due to its wide range of capabilities, features, and characteristics. The most common features (of it) are highly beneficial for the businesses to save substantial amount of money and valuable time along with ensuring top-level security and reliability. The main features and characteristics of blockchain technology are summarized as follows [138–140]:

- A peer-to-peer network of diverse nodes that check, verify, share, and store the transactional activities on the network.
- Referred to as a digital distributed ledger of immutable records of assets.
- Entire verified record is stored in block of information, which is shared with all nodes of the network simultaneously.
- Each node participates in verification and authentication process of a transaction through different techniques to build consensus among the nodes such as:

  - Proof of Work (PoW)
  - Proof of Stake (PoS).

- At the core of the blockchain technology stands three main technologies such as:

  - A source of computer resources to process and store database of transactions

- Distributed peer-to-peer network technology with shared ledger
- Cryptographic keys used to encrypt the data storage in the ledger.

- This technology uses two types of cryptographic keys known as private and public keys.
- One user of the technology is defined by both of the private and public keys jointly referred to as "digital signature" or digital identity. This is the basic component used for verification and authorization of the transactions on the network.
- Numerous types of assets, both tangible and intangible can be traded and managed through this secure technology of distributed ledger.
- It reduces the cost of ledger management and transaction management in any kinds of business and removes the barrier and charges of using the services of third-party or government agencies for their approvals and agreements.
- It is highly secure and immutable technology, which cannot be tempered. Even if some mistakes occur, it is picked up by numerous nodes and the correction is stored in the form of another block of data pertaining to that error. Thus, it is highly reliable and provides a complete details of and trail of the activities transparently.
- It can be used in a range of applications such as cryptocurrency, voting systems, government data management such as lands, revenue, contracts, purchases, finances, banking and insurance, defense and security, and many others.
- "Smart Contracts" is another futuristic use of this technology to automatically manage the entire process, activities, transactions, modification, changes, and other aspects of a business contract based on well-defined rules through this technology.

In the nutshell, blockchain technology is the future of modern processes in almost all domains of industries, businesses, governments, and societal management systems.

History of Blockchain Technology

The traces of blockchain technologies can be tracked long before it came into existence in 2008. The first trace of this technology can be found in the work of David Chaum, who enunciated a mechanism or protocol, which is like blockchain technology in 1982 in his research dissertation named as "*Computer Systems Established, Maintained, and Trusted by Mutually Suspicious Groups*". The future based on this concept was accomplished by Stuart Haber and W Scott Stornetta in 1991. Their further work along with Dave Bayer resulted in the incorporation of Merkle Tree into the design of the secure block of data through untampered timestamp. Thus, the concept of secure block of data in which timestamp cannot be tempered started getting materialized in those days.

Fast forwarding from 1992 to 2008, a group of people or just a person, who was popularized as Satoshi Nakamoto introduced highly secure and improved block of decentralized database with the help of hashcash-like method to timestamp the digital transaction without any intervention of any third-party in the process. This improved version of the blockchain technology was primarily coined as block and chain denoted separately but later on it was merged. This is very important to note that Satoshi Nakamoto remains unknown and invisible physically in the field of this unknown technology other than just one clue that this group or a single person belonged to Japan (at the time of writing this book). The chronicle timeline after this development is summarized in the following points.

- **2009**—The launching of the first-ever cryptocurrency based on the improved blockchain technology developed by Satoshi Nakamoto. The bitcoin became the new introduction of blockchain technology, which then entered into numerous other industries and business domains in the modern digital world.
- **2010**—The first commercial transaction of bitcoin based on blockchain technology took place on May 22, 2010 when Laszlo Hanyecz bought two pizzas for 10,000 bitcoins from local pizza vendor in Florida [141].
- **2011**—The second digital product based on blockchain technology and named as NameCoin DNS system was developed in April, 2011 and Litecoin later in October 2011.
- **2014**—The total size of bitcoin ledger file became 20 GB.
- **2016**—The previous structure of term was "Block Chain". This was merged to create new term known as "blockchain" technology
- **2020**—Total size of the file of bitcoin became over 200 GB (and, still expanding).

The use of blockchain technology has become a commonplace activity in numerous business processes across all industries of the world. Newer products and services are emerging on the technological sphere that are using blockchain technologies to make the most of it.

Major Terms Used in Blockchain Technology

Blockchain technology uses different terminologies, both technical and commercial, for its use and understanding. A few very common terms used to define and convey the information among the technical and commercial users are mentioned below:

Cryptographic Hash

Cryptographic hash is an output code of a message of arbitrary size produced in equal length of code through cryptographic hash function generating algorithm. The cryptographic hash converts data of any types, sizes, or formats into output text string of equal length consisting of numbers and letters. The output is always unique for any data hashed through hashing function algorithm [142].

The main features and characteristics of cryptographic hash function used in blockchain technology are mentioned below:

- Hash code is easy to create with the help of an algorithm and almost impossible to break.
- It is not understandable though you can read the characters—letters and numbers used in the message but no meaning can be achieved.
- The cryptographic mechanism has different types, sizes, and formats of input data but it encrypts into a same-sized output message based on the technology used in hashing algorithm.
- The basic function of hashing is to compress the input message into a smaller size and cryptographic process means the encoding of the entire message. Thus, the encoded text for the compressed input message is the meaning of cryptographic hash.
- The encrypted hash code acts like a fingerprint of the file encoded with the cryptography.
- A minor change in the input message alters the output cryptographic hash message hugely that is known as Avalanche Effect.
- Cryptographic hashing is referred to as non-reversible or you cannot break and get the input message from the output code. The hackers either try to build their own huge database of hash codes and run comparison of the codes with huge database to try to break into the code or unleash brute force attacks (i.e., trying all possible options with repeated trials) for the same purpose.
- Creation of similar hash code is almost impossible due to the ability of the cryptographic hash of message collision resistance.
- Cryptographic hashing is based on mathematical procedure to generate unique output strings of data through a mathematical algorithm.
- Cryptographic hashing is also known for its foreordain property, which means that the output code for the same message is always the same.
- The main domains of application of this hashing function include:

  - Duplication detection or unique file detection applications
  - Fingerprinting and password verification processes
  - Digital signature applications
  - Proof of Work (PoW) and Proof of Stake (PoS).

- The most common cryptographic hash algorithms used in the modern technological sphere include MD5, SHA-1, Bcrypt, Whirlpool, AES, and others.

Transaction

A transaction in blockchain technology is a procedure of activities to send message/data from one user account to the other concerned user accounts. It involves numerous steps to complete a transaction, which is later registered in a block of data in the blockchain database. The main steps that constitute a traction in blockchain include [143]:

- The request for a transaction originates from a single end-user in the blockchain network
- An authentication process starts for that particular request before further-proceeding
- A block is created representing that particular transaction
- All nodes of the blockchain nodes get the copy of that transaction block
- Each node has to validate that transaction based on certain validation procedure
- The nodes are rewarded for their work of validation know as Proof of Work (PoW)
- The verified block is added to the existing chain of blocks
- The update is sent out across the network to all nodes in the blockchain network
- This brings the completion of a transaction in blockchain technology.

Proof of Work

Proof of Work, precisely referred to as PoW in blockchain technology, is a process of developing consensus on the validation of a transaction and mining of a new blockchain token. In other words, it is a mathematical puzzle validation in blockchain network to avoid any malicious use of activities in the network. The main characteristics of "Proof of Work" are mentioned below [144]:

- A decentralized consensus procedure in which the members of the blockchain network expend efforts or work for solving an arbitrary mathematical puzzle for avoiding any kinds of misuse or malicious use of the network.
- This process is extensively used for cryptocurrency mining and transaction validation.
- This automatic mechanism plays the role of third-party verification agents used in the traditional systems of transactions.
- Due to extensive usage of power on computing for solving the puzzle, PoW is being replaced by a new method of consensus known as Proof of Stake (PoS).
- This is also known as the proof of expending the computational resource or power in the network for achieving the consensus on the work performed by a node.

Block

Block is a kind of data structure used in the blockchain technology to store the entire details of a transaction in it. The stored data in the block is properly authenticated, validated, and encrypted for maintaining a high level of security and reliability. The most common characteristics of a block used in the blockchain technology include [145]:

- A block is a permanent storage container of data of a validated transaction, which cannot be altered or removed once it is closed.
- A block contains the encrypted data from the previous blocks and the future blocks to make it more robust and reliable.
- The new block is created only after the validation of the information in a particular block by developing consensus among the nodes.
- Blocks of information can be used for a range of transactional information including cryptocurrency, governmental activities, revenue transition, and many others.

Mining

Mining is a process that is used in cryptocurrencies to either generate the new coins or to validate the transaction over the blockchain network. A huge network of decentralized nodes spread across the globe are the main source of mining that solve the mathematical puzzles for generating a new value on the cryptocurrency network or the validation of the transactions that are taking place on the networks powered by the blockchain technology. The blockchain technology offers an option of rewarding the mining nodes for their work with the proof that they expended the computation resources or power for the purpose of solving the mathematical puzzles.

Timestamp

The fingerprint or authorized code against a message that may be a value, title, text, or a digital asset when verified and stored in the block is known as the timestamp. This timestamp is also known as hash in the blockchain technology. Hash is a unique code generated against an arbitrary input of different size, type, and content of a digital file. Initially, the blockchain technology was invented for time-stamping purposes. A few very common characteristics of timestamp and timestamping process used in blockchain include [146]:

- The encoded hash in equal size and unique code (together) is known as timestamp of the data.
- The data may be a text file, a value, a string of text and numbers, or any title or a digital data file.
- All features of timestamp are similar to hash or fingerprint, which holds a huge amount of information about the transaction or activities within it, which cannot be removed or altered.

Stack of Technologies Forming Blockchain

Blockchain technology used in different domains is a combination of software and hardware technologies, which constitute a robust network of blockchains that is reliable, secure, and encrypted. The most common technologies that form blockchain stack are mentioned below.

Cryptographic Keys

Cryptographic keys are two different types of keys or codes used in building digital signature for authentication of the ownership of the transactional data in a blockchain-based network. There are two types of cryptographic keys such as [147]:

- Public key
- Private key.

A public key is a code that is used to identify the address of a user-entity in the blockchain technology. It is also paired with the private key for establishing ownership on the assets associated with the private key or address. It is like the email address used for email communication. Anyone can send a digitally signed transaction to your public key but cannot access it without having the combination of the private key. The public key is also encrypted form of message or information, which can easily be encoded for any transaction on blockchain technology but it is very difficult to decode or decrypt to find out the original message in the key.

On the other hand, the private key is very vital for the ownership, security, and control over the blockchain assets. It is a long string of digits, letters, or combination of both. It can be a huge number in the form of the following expressions:

- A long binary code of 256 characters
- Hexadecimal code of 64 digits
- A long mnemonic phrases
- QR (Quick Response) code.

The main features and functions of a private key in blockchain technology are mentioned in the following summarized points:

- It is a long, randomly-generated, and encrypted string of numbers or alpha-numeric characters.
- Used for the authentication and ownership of the assets on blockchain.
- Private keys are normally very long strings in such a way that they cannot be cracked through brute force attacks.
- Creating a public address or public key from a private key is very easy and possible through a mathematical calculation or mathematical puzzle solution but the reverse engineering for decrypting private key is almost impossible.

- It should always be stored carefully and should never be shared with anyone. If it is lost once, you would lose all assets on blockchain technology.

Peer-to-Peer Network with Shared Ledger

A peer-to-peer network with shared ledger is a network of wide range of nodes or computer machines installed with the blockchain technology that enables them to coordinate in a peer-to-peer network environment without any intervention of a server or any other network agent or any other third-party—either human or technological. A shared ledger of transactions operates and updates on all peer nodes across the network. This means, all nodes have the same shared ledger, which is updated on a regular basis through a robust process of authentication and authorization.

Computing Resources to Store Transactions and Network Records

The computing resources that are used for storing the network transactions on a blockchain technological network are those computing nodes that are private entities added into the global network of any private blockchain service or any private blockchain service. Those computing resources are hosted by the interested parties and they pay for the expenditures and operational charges. The owners of those computing resources to store blockchain records are shared ledger, which are verified and mined by that node. For that mining or verification purposes, the owners of those nodes are rewarded with associated cryptocurrency assets such as bitcoin or any other digital currency or any other services of tangible or intangible assets on the blockchain.

How Does Blockchain Technology Work?

A blockchain technology-based network consists of three main entities named as **Node, Blocks,** and **Miners**. The entire process of working is based on the activities among and on those elements of the network. The working functions and roles of the three elements are described below separately [148].

Node

A node in this case is a computer machine that hosts the mining software, unique alphanumeric ID number, and distributed ledger of blockchain technology. It is not a property of any particular controlling party but anyone has right to add a node to the blockchain node by fulfilling the requirements of the network and technology stacks. The node is rewarded for its work of mining and authenticating the transactions on the blockchain network automatically without any intervention of any centralized authority or third-party organization.

Block

A block is the basic container that contains the encoded information or message including the transactional details in the chain. A block consists of the following three major elements:

- **Data**—This is the information in a block, which is encoded, authenticated, and verified with additional details regarding the past and future blocks in the transactional series of activities on the blockchain network.
- **Nonce**—A nonce is randomly created whole number of 32-bits. It is generated when the block is created by the network. The most important functionality of nonce is to generate the block header's hash. The hash is interlinked with the nonce number. The hash is a long string of zeros with a very small value. It consists of 256-bit number, mostly containing zeros in the beginning of the string.
- **Miners**—The miners are the computing resources attached with a node and unique ID number working in relationship with the blockchain technology software. The main responsibility of miners is mining of the new encrypted resources and authentication and verification of the transactions in the network. The miners generate their own block of information that consists of a nonce of 32-bits and a long string of zeros with a very small value (automatically generated numbers) linked with each other. A new block that is added to the chain of blocks after mining or verification and validation consists of many things such as tile hash, nonce, reference to previous hash in the chain, and others. Miners find out the matching combination between the 32-bit nonce and the hash of 256-bit code through a mathematical algorithm running in a puzzle solution application. There will be around four billion possible nonce-hash combinations to solve to reach the golden nonce. The golden nonce is rewarded with a certain value of that asset defined on the blockchain asset network.

What Is Distributed Ledger Technology (DLT)?

Distributed Ledger Technology, precisely referred to as DLT, is a detailed structure of entire stack of blockchain technology protocols that enable the entire ecosystem of blockchain technology to create, authenticate, verify, and store the message or information in the encrypted form on a distributed ledger (of long queues of blocks of information located on every node in the network). This is the core spirit of the blockchain technology or it is the real philosophy of blockchain distributed ledger technology commonly adopted in the modern digital asset management as well as in all types of cryptocurrencies and Ethereum platforms [149].

Types of Blockchain Technology

Blockchain technology has become very popular within a few years in all domains of businesses, governments, and social sectors due to the security, privacy, and robustness that it offers to the organizations and people. This technology also offers flexibility to customize the use of this technology in different ways. The most common categories in which the blockchain technology can effectively be used are the following [150]:

- Permissionless blockchain category
- Permissioned blockchain category
- Hybrid features of the above (both) categories.

In the permissionless category of blockchain, it is used in the way that no permission or control is established over the nodes of the network. The nodes in this category can join without any permission with a pseudo-anonymous approach. There is no centralized authority to regulate or control the network types falling under this ambit. The main type that works on the basis of this category of blockchain is: Public blockchain network.

In the permissioned category of blockchain, there is a centralized control of a group, consortium, or an authority. In this category, the nodes are granted permissions by the centralized authority thus, reducing the security-level, privacy, and other major features of the traditional blockchain technology. The types of blockchain networks falling under the permissioned category of networks include: Private blockchain network and Consortium blockchain network.

The third category of blockchain technology is the middle-way of the above-mentioned main categories (permissioned and permissionless). This uses the combination of the features of both categories. An example is the Hybrid blockchain network.

The details of all those types of blockchain technology networks are described separately in the following sub-sections. Meanwhile, this is very important to note that the core objectives of developing blockchain technology were to establish a

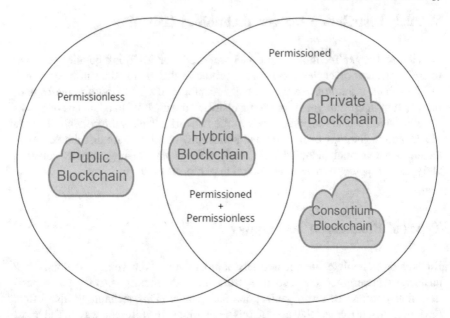

Fig. 4.1 Schematic diagram of categories and types of blockchain networks

network that has no influence of any centralized controlling authority or entity and to maintain high-level of privacy, anonymity, security, and reliability. If there is a control on the entry of the nodes in a network, the network gets less secure due to lesser number of nodes in a blockchain network to process the authenticity of the transactions. The example of such networks are the private and consortium networks. Let us know about these types of blockchain networks with more details.

Figure 4.1, it is very clear that the types of networks falling in the permissionless category are the public networks with full features and capabilities of blockchain network while the permissioned networks are controlled networks and the hybrid one shares the features of both the categories.

Public Blockchain

A public blockchain technology is characterized by the most common features that were the foundation of the technology such as:

- A sub-set of permissionless category of blockchain
- Offers full decentralization of the control
- No third-party is involved at any level of control
- Any public node can participate in the network with basic prerequisites
- All nodes have equal rights in the network to access and work in the blockchain network

- All nodes can create new blocks, mine data, and authenticate transactions
- The examples include bitcoin, Ethereum, Litecoin, and so on.

Private Blockchain

The private blockchain is a controlled type of blockchain technology that is centrally managed by a third-party organization or an entity. The main features of this type of blockchain include:

- A type of permissioned category of blockchain technology, which is fully managed by the third-party organizations or persons.
- The centralized controlling authority of private network is one entity
- Any new node to be added to the network is fully under the authority of the network management body
- All nodes do not have equal rights delegated by the central management system
- Public access to participate in the network is not allowed
- The examples of private blockchain networks include Hyperledger, Ripple, and so on.

Consortium Blockchain

A consortium blockchain, as the name implies, is a network that is private-like network but with multiple controlling authorities grouped in a consortium. The main features and characteristics of consortium blockchain are:

- It is a type of blockchain network that falls under permissioned category
- The controlling authorities are more than one or a group of companies known as consortium
- It is more decentralized in nature than the private networks, which are fully managed and controlled
- More secure than public blockchain networks but less secure than private blockchain networks
- The examples of consortium blockchain technology include CargoSmart, and R3 software consortium platform.

Hybrid Blockchains

The hybrid type of blockchain technology has the characteristics and features of both the private and public blockchains. The main features of this type include:

- Blockchains that are managed by one organization with limited features of private network and other features of a public network for maintaining higher level of security.
- This is a domain-specific performance improving solution based on certain customized features and capabilities.
- The example of hybrid blockchain type of technology include the IBM Food Trust designed for the improvement of the food supply chain's performance while maintaining the high-level of security and privacy of the network.

Typical Uses of Blockchain Technology

The global market size of blockchain technology was just USD $1.5 billion in 2018. According to the Statista information, this market size is projected to cross USD $162.84 billion by 2027 with an exponential growth due to the high level of traction achieved by this technology in all domains of businesses, governments, and other fields of our day-to-day lives [151].

The use of blockchain technology is spreading across the domains, sectors, and fields all around the world. Among the use cases of blockchain technology in different fields, a few very typical ones are mentioned below with more details.

Cryptocurrency

Cryptocurrency was the most prominent use case of blockchain technology, which utilizes the power of blockchain technology to its fullest in the public category of this technology with full anonymity, security, and privacy. The popularity of cryptocurrencies, especially bitcoin, overshadowed the name of this basic technology in such a way that blockchain technology looks (sometimes) like a sub-name of bitcoin in some normal audience in the world.

According to Research and Markets projections, the global market size of cryptocurrencies will reach a whopping USD $32.42 by 2027 from just USD $1.78 billion in 2021 with a staggering growth of over 58.4% CAGR (Compound Annual Growth Rate) during the projected period. These figures are quite clear indicators of the level of traction of blockchain in the cryptocurrency market [152].

The first cryptocurrency named as bitcoin was created in 2009. Within just over a decade, there are more than 18,000 cryptocurrencies that exist in the global market size. Many new cryptocurrencies are in the development. A large number of governments are also considering to introduce their own cryptocurrencies to benefit from the huge market potential of blockchain technology in the domain of global payment systems. The most common or most popular cryptocurrencies in 2022 are bitcoin (BTC), Ethereum (ETH), Litecoin (LTC), Cardano (ADA), Dogecoin (DOGE), Binance Coin (BNB), Stellar (XLM), Dollar Coin (USDC), and so on.

Non-Fungible Token (NFT)

Non-fungible tokens, precisely referred to as NFTs, are a type of digital assets encrypted in the blockchain technology as an asset for trading like a commodity (unlike a cryptocurrency). It is a unique ID and asset created by one owner of that data. These assets cannot be exchanged or interchanged with the other commodities. In other words, NFTs are types of cryptographic assets created in the blockchain technology with unique identity number or code and metadata that differentiate this data from the other similar kinds of data on the blockchain asset network. The main features and characteristics of NFTs are listed below [153]:

- Unique code or ID on blockchain that cannot be altered or replicated
- NFTs are real-world items and assets such as photos, arts, real-estate, properties, and others
- Blockchain facilitates secure and transparent trading of unique assets created in the forms of NFTs that replace the real-world items
- This is a great area of trading and business of copyrighted assets, tangible assets, and digital assets like music, videos, photos, arts, training content, and may others.

According to the Research and Markets projections regarding the NFTs, the global market size of NFTs will reach USD $82.43 billion by 2026 from just USD $21.33 billion in 2022 with a whopping growth of over 40.2% CAGR [154]. The most common example of NFTs is the use of Ethereum, which acts as both NFT and cryptocurrency simultaneously in different types of trades.

Smart Contracts

As the name implies, smart contracts are the (automated) contracts that run automatically by the completion of different activities, actions, functions, and other responsibilities defined in an agreement after the verification and authentication of the same without any involvement of the third-party. Smart contracts have become so popular in modern processes of agreement or contract automation in all types of projects in a wide range of domains in the industries [155].

According to the Verified Market Research projections, the global market size of smart contracts based on the blockchain technologies will reach to over USD $770 million by 2028 from just USD $145 million in 2020 with a gigantic growth of over 24.55% CAGR during the forecast period between 2021 and 2028 [156].

The most important reasons for the businesses to choose blockchain technology for smart contract applications include the highest level of security of the transactions taken on the blockchain due to capability of consensus building, immutability of transaction reversal and alteration, capability of technology to avert any kinds of replication or duplication, and many more. These contracts can incorporate all major phases of a project, and many types of other contracts such as escalation matrix, payment matrix, service level agreements (SLAs), responsibility matrix, and so on. All functions and activities such as task accomplishment, collaboration, asset sharing, billing, operations, and many others required for a particular project are automatically accomplished and acknowledged by the roles that have been designated to complete those tasks. The release of payments and all other final financial transactions are also completed automatically without any intervention of a third-party entity.

Financial Markets

Financial markets, especially in the modern banking and other financial services, the use of blockchain has become very fundamental. The growth of the size of the financial market has increased significantly during past few years. According to the forecast of Statista in 2020, the global market size of blockchain-powered banking and financial services would reach USD $22.46 billion by 2028 from just USD $0.28 billion in 2018 with a dramatic exponential growth during the projected period and even beyond that period [157].

The most promising areas for the use of blockchain technology in the global as well as local financial markets include:

- International financial transaction management and securities
- Mainstream financial settlements and payments both locally and internationally
- Clearing, securities, and insurance services
- Financial derivative management and trading
- Corporate governance and credit bureau management
- Micro-payment management and financial services at local levels
- Management of pre- & post-trade processes
- Assets management, data registries, and repository services
- Enterprise resource management and financial process automation.

Electronic Voting

Electronic voting is another major use case of blockchain technology due to its capabilities of immutability, duplication- and replication-free, traceability, transparency, and security. There are many countries that have either conducted electronic voting through blockchain technology or are developing such systems with pilot projects. The most important countries that have already implemented electronic voting through blockchain technology at the local levels include the USA, Russia, Sierra Leone, and Japan (so far, at the time of writing this book). Sierra Leone is the first country that has conducted voting through blockchain. The other major countries that are also working on different levels of projects of voting systems based on blockchain technology include India, South Korea, Thailand, and others [158].

Record Maintenance

Maintenance of records transparently without any duplication, fabrication, and alteration is another very important domain of application of blockchain technologies. This is a very vast field, which covers the record keeping in different departments of government such as land, properties, reports, researches, history, and many other assets—both tangible and non-tangible. Meanwhile, numerous private and industrial domains have also huge prospects in using blockchain technology for the management and maintenance of records in different processes, activities, transactions, histories, and many more with highly effective ways.

Supply Chain

According to the Allied Market Research projections, the global market size of blockchain-based supply chain is expected to cross USD \$9.853 billion by 2025 from just USD \$0.093 billion in 2017 with a huge growth of over 80.2% during the projected period between 2018 and 2025 [159].

There are many companies and countries who are using a range of supply chain systems powered by blockchain technology from major players in the technological domain such as IBM and others. The most common supply chain areas that have already proved to be suitable include the vaccination supply system, food supply chains, digital identity verification and purchase systems, and container logistics. It is expected that many other domains will be explored in the near future where the use of blockchain technology for the management of supply chains will prove highly effective [159, 160].

Government

Government is one of the most promising sectors for the use of blockchain technology. This is because the procedures, assets, information, orders, decisions, and researches, discussions, and stakeholders in this sector are highly sensitive. Also, these should be treated with high level of security, privacy, and secrecy with deep traceability of the events and transactions for any kinds of future requirements to maintain transparency and confidence of the public in governments.

There are a wide range of areas and departments in the government sector where the blockchain can be deployed with full confidence. A few of those main areas and activities of governments are listed below:

- Electronic voting systems
- Land property management
- Licensing and penalty systems

- Judiciary and legislation systems
- Utility management and provisioning systems
- Social securities, insurance, and other financial systems
- Banking and trading systems
- Defense and security research and development
- Cryptocurrency and digital asset management.

In a nutshell, the government sector is the biggest one that can use the potential of blockchain technology in wide range of its activities, processes, departments, and domains to make them more robust, secure, reliable, transparent, and effective for their respective people.

Impact of Blockchain Technology on Cybersecurity

Blockchain technology is very well known for its capabilities of providing high-level security, end-to-end encryption, anonymity, immutability, data integrity, confidentiality, and privacy. If all those factors are related to one way or the other, they associate with the robust security of data, communication, applications, and devices. Simply put, the impact of blockchain on cybersecurity is highly desirable and even prospective.

Cybersecurity is based on the CIA (Confidentiality, integrity and availability) triad model. This model helps the cyberworld in protecting valuable data from getting damaged or stolen, safeguarding edge devices or network elements from any kinds of external or internal malicious exploitation, and maintaining the performance and working efficiency of the cyber-based systems of communication, data management, process management and others. All those functions fall under the summary of CIA triad model, which expresses three major components such as [161, 162]:

- Integrity
- Confidentiality
- Availability.

Let us explore the impact of blockchain technology on the cybersecurity by comparing the main features and capabilities of blockchain technology and finding their use in the enhancement of the major security issues in the cybersecurity domain. This will provide a better overview of the impact of blockchain technology on the cybersecurity enhancement. The main risk areas of cybersecurity management include the following:

- **Exploitation of protocol vulnerabilities**—This is one of the major areas that is exploited by the hackers to intrude into the cybersecurity or security systems of a network or an environment such as home automation, local area network, access control, and others. There are numerous communication protocols that have vulnerabilities that are exploited by the hackers. If all communication protocols are powered by the secure capabilities of blockchain, they can provide higher level of security through secure communication, authentication, verification, and encryption of the data during the communication and transmission.

- **Edge device vulnerabilities**—Edge-devices are the most preferred nodes or elements for the hackers or malicious users due to numerous ways to exploit them and their vulnerabilities. By using the power of blockchain technology, cybersecurity personnel can decentralize the management and control through blockchain. In fact, highly secure authentication, authorization, and verification processes could be handled by the basic features of blockchain technology.
- **DNS systems breaches**—The breach of hierarchical systems of Domain Name Services (DNS) are another main target of the malicious users. If the DNS system is decentralized and controlled through extremely secure, reliable, and robust system of blockchain for managing the DNS services, a major domain of exploitation or vulnerability can be controlled or eliminated while maintaining the performance of the system simultaneously. By using the power of blockchain, Distributed Denial of Service (DDOS) attacks can easily be averted or reduced significantly and security can be enhanced tremendously.
- **Internet of Things (IoT)**—IoT is a vast domain of concern for the cybersecurity nowadays. It consists of a huge number of devices, which are run by different operating systems, firmware, controlling systems, connectivity vulnerabilities, and many others. All those factors can easily be exploited by the hackers to intrude into the system and breach the data or security of the network. If the decentralized management of all those devices connected to the IoT environment are managed by the decentralized processes powered by the blockchain technology, a huge boost in the cybersecurity field can easily be achieved.
- **Data integrity breaches**—The breach of data integrity can happen either on the storage device or in the transition. The end-to-end encryption and decentralized control of the transactions during the communication powered by the blockchain technology, can increase the security of data integrity significantly.
- **Miscellaneous issues**—Numerous other issues such as patch installation, verification, access control, and many other monitoring and risk assessment issues can easily be powered by the highly secure features of blockchain technology (to enhance the security level hugely).

Thus, the impact of the blockchain on cybersecurity is highly desirable and many companies, organizations, and governments are incorporating the power of this technology for boosting cybersecurity of their respective systems. This trend is expected to grow further in the coming days.

Sample Questions and Answers

Q1. Write down at least three main features and characteristics of blockchain technology.

A1. Three main features and characteristics of blockchain technology are:

- A peer-to-peer network of diverse nodes that check, verify, share, and store the transactional activities on the network.
- Referred to as a digital distributed ledger of immutable records of assets.
- Entire verified record is stored in block of information, which is shared with all nodes of the network simultaneously.

Q2. Define Cryptographic Hash. What does it do?

A2. Cryptographic hash is an output code of a message of arbitrary size produced in equal length of code through cryptographic hash function generating algorithm. The cryptographic hash converts data of any types, sizes, or formats into output text string of equal length consisting of numbers and letters. The output is always unique for any data hashed through hashing function algorithm.

Q3. What are the main characteristics of "Proof of Work"?

A3. The main characteristics of "Proof of Work" are mentioned below:

- A decentralized consensus procedure in which the members of the blockchain network expend efforts or work for solving an arbitrary mathematical puzzle for avoiding any kinds of misuse or malicious use of the network.
- This process is extensively used for cryptocurrency mining and transaction validation.
- This automatic mechanism plays the role of third-party verification agents used in the traditional systems of transactions.
- Due to extensive usage of power on computing for solving the puzzle, PoW is being replaced by a new method of consensus known as Proof of Stake (PoS).
- This is also known as the proof of expending the computational resource or power in the network for achieving the consensus on the work performed by a node.

Q4. What are the main two types of cryptographic keys?

A4.. There are two types of cryptographic keys:

- Public key
- Private key.

Q5. What do you mean by NFT (Non-Fungible Token)?

A5. Non-Fungible Tokens, precisely referred to as NFTs, are a type of digital assets encrypted in the blockchain technology as an asset for trading like a commodity (unlike a cryptocurrency). It is a unique ID and asset created by one owner of that data. These assets cannot be exchanged or interchanged with the other commodities.

In other words, NFTs are types of cryptographic assets created in the blockchain technology with unique identity number or code and metadata that differentiate this data from the other similar kinds of data on the blockchain asset network.

Test Questions

1. How does Blockchain Technology work?
2. What is Cryptographic Hash?
3. How does Distributed Ledger Technology (DLT) work?
4. What are some typical uses of blockchain technology?
5. What is the impact of blockchain technology on cybersecurity?
6. Why is the supply chain important?
7. What makes record maintenance so important?
8. Describe hybrid blockchains.
9. How do you define a 'block'?
10. How do computing resources store transactions and network records?

Chapter 5
5th Generation Wireless Technology

An Introduction to 5G Technology

The fifth-generation technology, precisely referred to as 5G technology, is a wireless technology based on the cellular cells like the existing 4G LTE (Long-Term Evolution) technology network. This technology has been designed for higher speed, performance, and efficiency in communication network. This is the advanced generation of wireless communication that offers seamless mobility with greater throughput and better utilization of the available wireless and other technological resources used in the materialization of this prospective technology.

There are numerous features, capabilities, and characteristics that make this technology as one of the most futuristic technologies for many years to come. A few very prominent features and characteristics of the 5G technology are summarized below [163–165]:

- It is the latest generation of wireless-based cellular technology developed under the auspice of 3rd Generation Partnership Project (3GPP), which consists of a huge group of numerous technological companies and related organizations that develop advanced wireless communication standards for the advancement of wireless technology.

K. Thakur et al., *Emerging ICT Technologies and Cybersecurity*,
https://doi.org/10.1007/978-3-031-27765-8_5

- 5G technology offers the capabilities to connect every device, people, objects, and machines seamlessly with full mobility and greater data speed to bring forth the fourth industrial revolutions in the world.
- It offers higher peak data speed in multi-Gbps, highly reduced network latency, gigantic capacity of network, greater reliability and availability, huge coverage area, larger number of users, and highly enhanced user experience to the end users.
- 5G technology is also known as heterogeneous network that works on the existing 4G LTE networks by expanding the effectiveness of resource usage with the help of the latest air interface and service layer protocols.
- It uses Orthogonal Frequency Division Multiplexing (OFDM) for signal modulation or encoding across the wide range of channels for reducing the interference between the signals.
- The other technology that improves the speed, performance, and latency of the network is 5G NR air interface, which is also known as "5G New Radio" interface.
- 5G technology is capable of using the combination of multiple spectrums of frequencies, commonly low-band, mid-band, and high-band frequencies simultaneously to increase the capacity of the network and data transmission.
- It operates in 6 GHz bands as well as in 24 GHz bands simultaneously by binding multiple channels of 20 MHz each to form bigger airwave for the transmission of bigger data rates and for the substantial decrease in network latency.
- 5G technology uses massive MIMO (Multiple Input, Multiple Output) antenna technology
- In other words, the 5G technology is capable enough to support multi-bands such as low, mid, and high as well as all types of spectrums such as shared spectrums, licensed and unlicensed spectrums of frequencies simultaneously.
- The designed capacity of 5G network is 100 times the increase in capacity of the network as well as network efficiency.
- It can provide about 20Gbps peak data speed and more than 100 Mbps average data transmission rate.
- The basic working principle and structure of base-transceiver station (BTS) is almost the same. The area around the base station is divided into sectors which are targeted to provide better airwave for greater data transmission.
- 5G wireless technology supports 3 channels of 100 MHz to combine in low and mid bands and up to 8 channels of 100 MHz to form 800 MHz combined band for data usage to increase the speed multiple times as compared to the LTE 4G technology. In fact, the latter combines 7 channels of 20 MHz capacity to form 140 MHz spectrum for data transmission.
- For combining, splitting, and sharing of the frequency channels, both 4G and 5G use Dynamic Spectrum Sharing (DSS) protocol.

Importance of 5G Technology

5G wireless technology is the next-generation technology, which is going to impact all spheres of life, business, society, governments, industries, procedures, and a range of related technologies simultaneously one way or the other. This huge impact on numerous domains is due to the power of 5G technology characterized by high-speed, adoptability, low-latency, enormous capacity, improved performance, greater efficiency, reduced carbon footprints, super-smooth mobility, and many others. All those features, characteristics, and capabilities of this next-generation wireless cellular technology make it highly important for all above-mentioned domains. Let us explore the importance of 5G with respect to a few major domains [166, 167].

- **Automation**—5G technology accelerates the process of automation in a range of fields such as home automation, industrial automation, office automation, business process automation, cybersecurity management, and many others by leveraging the high-speed, huge capacity, and reduced latency in the network response.
- **Governments**—The importance of 5G for the governments is very huge in different departments and sectors such as land-records, smart cities, utility management, virtual monitoring, defense and security, voting systems, and so on. Thus, 5G can help governments exploit the available potential in all government sectors substantially—via wireless connectivity.
- **Social life**—In the social life, 5G enhances the standards of life by creating a great level of user experience in communication, web surfing, service utilization, software-driven services, telemedicine, well-being, infrastructure improvement and many others.
- **Modern technologies**—The most futuristic technologies such as IoT, cloud computing, software defined networks (SDN), network function virtualization (NFV), and many other emerging technologies can only get materialized if the communication bandwidth is faster, with greater capacity, low-latency, and better efficiency. All those features are provided by the 5G technology to realize those modern technologies.
- **Cybersecurity**—Though 5G technology can pose a few major challenges for the cybersecurity personnel to maintain the high-level security of an enormous network of billions of devices connected through a range of technologies and domains, it offers better opportunity for the security experts to leverage the power of 5G technology to develop faster and effective solutions that can monitor and avert the emerging cybersecurity threats in the real-time environment.
- **Industry 4.0**—The fourth industrial revolution, precisely referred to as 4IR or industry 4.0, is the fourth revolution in the world industry powered by the heavy automation, artificial intelligence, cloud computing, cognitive computing, cyber-physical systems (CPS), industrial internet of things (IIoT) and many others. All those modern industrial concepts cannot produce the desired objectives until a power network with greater capacity, speed, performance, and latency is available to use. The 5th generation wireless network comes into play to fill that gap in both last mile as well as the core levels of communication networks.

- **Remote workplace**—With the advent of 5G technology, the domain of remote workplace will expand exponentially by using a range of heavy platforms to work in the real-time work environment powered video, audio, and other effective means of communication, collaboration, and coordination like a global office or workspace.
- **Healthcare**—5G can improve the performance or remote healthcare services and applications significantly. The remote medicine will get more benefits from the emergence of next-generation (5G) technology. E-Healthcare can be greatly benefited.

In a nutshell, the fifth-generation wireless technology will revolutionize all processes, applications, businesses, industries, governments, and social activities tremendously.

Evolution of Cellular Networks

Cellular networks have evolved through 4 decades out of numerous analog as well as digital technologies with task-specific applications and communication protocols. Roughly, the start of the cellular wireless technology dates back to 1980s when formal deployment of a cell-based communication system was done. The subsequent events, ages, and generations of cellar technologies are summarized in the following categories [168].

First Generation (1G)

The 1st generation cellular technologies can be tracked back in 1960s with very primitive research and development that pave for the starting mark of the first-generation technology of cellular communication. The main features of 1G technology include:

- The first-generation technologies started in 1979 when a commercial communication system based on cellular technology was launched in Tokyo, Japan.
- Later in 1980s, numerous countries launched the same in different names such as AMPS (USA), NMT (Nordic), and others.
- The 1st generation technologies used 800 MHz band for voice and signaling based on frequency division multiple access (FDMA) technology.
- This technology was analog wireless technology; so, highly prone to interference and numerous other disadvantages.
- The first-generation technology was limited to just voice calls with no support for text or any other software-based other applications.
- The initial paging networks are also considered as the part of first-generation wireless networks.

Second Generation (2G)

The second-generation cellular networks evolved to digital networks from the first generation (1G) analog networks. The frequency used by the second-generation networks was the same as used in the 1G networks. The main features and characteristics of the second-generation cellular wireless mobile technology are summarized as follow:

- Along with voice, the 2G technology supported text messaging, multimedia messages, and very low-speed Internet services simultaneously
- The operating frequencies of this technology were 1800 and 900 MHz bands
- It used the time division multiple access (TDMA) instead of frequency division multiple access (FDMA) modulation used in 1G analog technology
- For improving the Internet speed, additional technologies such as General Packet Radio Service (GPRS) and Enhanced Data Rate for GSM Evolution (EDGE) were introduced on the 2G networks powered by Global Systems for Mobile Communication, precisely known as GSM, which was launched in 1991 in Finland.

Third Generation (3G)

3G cellular technology is based on the core structure of wireless network governed by the Universal Mobile Telecommunication Systems (UMTS) standard for the core network to provide better Internet speed and performance. The other examples of 3G wireless technologies include IS95 (Interim Standard 95) used in the USA, CDMA-One, CDMA2000, and others. The most salient characteristics, capabilities, and features of 3G cellular technology include:

- The first role out of 3G technology-based networks was started in 1998.
- It offered better speeds up to 21.6 Mbps under HSPA + (Evolved High Speed Packet Access plus) technology stack enhancement. Initially, it was able to support up to 2 Mbps under different internet standards and technologies such as 1xRTT, 1XEv-DO, and others.
- At the core of the network, this technology started utilizing packet switching to improve the data throughput and performance of the data transmission as compared to the circuit switching used in the 2G and 1G networks at the core network.
- The operational frequency band of the 3G technology was 2100 MHz with channels bands of 15–20 MHz frequencies.

Fourth Generation (4G)

The fourth-generation cellular technology was developed under the consortium of 3rd Generation Partnership Project (3GPP) standard body. The most fundamental technology that refers to the evolution of the fourth-generation technology is known as Long Term Evolution (LTE). Meanwhile, the Wi-Max version is for fixed and mobile wireless communication. The other features and capabilities of 4G technology include:

- First launched in 2009 in Norway.
- It was released in two version referred to as R8 and R9.
- 4G uses Orthogonal Frequency Division Multiplexing (OFDM) as the core wireless modulation technology along with the support for multiple input and multiple output capabilities of wireless antenna for increased throughput.
- This technology mostly operates at 700 MHz band and supports high speed services based on video streaming, voice, and other multimedia systems.
- The average peak speed supported by the 4G networks was up to 50 Mbps and average speed ranging in between 15 and 20 Mbps in real-world environments.

Fifth Generation (5G)

Fifth-generation of wireless technology is considered as the futuristic technology, which is capable of providing about 100 times faster speed as compared to the predecessor (4G) technology. It has reduced latency and improved performance by utilizing the modern supportive wireless and core technologies to build this state-of-the-art technology. This technology is designed to support up to 10 Gbps Internet speed with very negligible latency resulting in highly desirable user experience and a range of service portfolios. The other information on 5G is described in different categories in this chapter with enough details and related factual information.

Sixth Generation (6G)

The sixth-generation (6G) technology is under development and research phase. This technology has not materialized yet. It can be characterized as an upcoming futuristic technology. The available details of this technology will be discussed in Chap. 10 of this book when talking about the futuristic technologies of the world in telecommunication and information technology (which is also related to cyberspace and cybersecurity).

Key Features and Capabilities of 5G Technology

The fifth-generation technology is known for its greater features, higher capabilities, bigger capacity, faster speeds, and low latencies. The most salient features associated with these categories of characteristics can open up new era of technological revolution in all domains and industries of the modern world. This can enable the interconnectivity of massive networks of networks consisting of different things such as cars, home appliances, sensors, industrial machines, office equipment, traffic signaling systems, drones, mobiles, and many others. This massive connectivity would lead to unprecedented automation in all businesses, processes, industries, and routine social and governmental activities. This technology can cater about 100 times more devices in a particular area of coverage as compared to the previous major wireless technology known as 4G technology. The speed of this technology can reach up to 100 times faster than 4G speed. The latency has reduced over 200 times as compared to the previous generation of technology. The most salient features, capabilities, and characteristics of 5G technology include [169, 170]:

- Basic characteristics include larger bandwidths, lower latency, and higher capacity.
- The 5G technology was launched in 2019 and will continue expanding for many years to come to reach at its full capacity and potential.
- The reduced latency can produce the response time to any visual event supported by the 5G technology is about 250 times faster than human response. Thus, a driver is about 250 times slower in response to any visual event during driving of a car.
- The speed of Internet supported by the 5G technology is about 10 times faster than the designated speed of 4G technology. 5G can produce as much as 10 Gbps. The bandwidth per unit area is about 1000 times in 5G as compared to the previous technology.
- This technology is specified under the specification of the next generation mobile network alliance (NGMN Alliance).
- The most salient capabilities can be defined as:

  - **eMBB**—Enhanced Mobile Broadband
  - **mMTC**—Massive Machine Type Communication
  - **URLLC**—Ultra-Reliable Low-Latency Communication.

- It uses millimeter waves, which are shorter than the microwave frequencies. Those frequencies increase the bandwidth and data speed significantly. The operating frequencies of 5G technology include 30 and 300 GHz bands.
- The use of high-frequency and millimeter wave-lengths result in higher bandwidth, speed, low latency, and greater capacity.
- It opens up the futuristic opportunities for numerous technologies such as Internet of Things (IoT), virtual reality and augmented reality (AR/VR), driverless vehicles, drones and robotics, and many others.

Architecture of 5G Network

The architecture of 5G wireless network technology is very modular and open powered by more software-defined infrastructure, and extensively open for custom development of interfaces, applications, and services. The architecture uses a few new components, application programming interfaces, cloud computing, edge computing, new radio interfaces, and many other incorporated into different domains of the network. The 3GPP specification describes the architecture of 5G technology into three major components as mentioned below:

- Core transport network
- Radio access network
- Service capabilities.

The complete architecture of 5G network consists of numerous parts of the existing 4G technology infrastructure along with the new additional advanced infrastructure that is fundamental part of the new technology. Thus, the entire architecture can be categorized in different factors as mentioned in below. The technical architecture of 5G networks is shown in Fig. 5.1 with a schematic diagram with different planes of services.

The core network consists of the servers, data storages, switches, and security systems sitting at the center of the service processing environment and backhaul transmission networks that interconnect with the core data network and radio access network through a range of specified connections and interfaces. The control layer, commonly referred to as Multi-Access Edge Control (MEC), shown in Fig. 5.1 is the control section of the services that interconnects the radio access network and core network functionalities through different control protocols and software defined services. Different control interconnection and interfaces are defined between the core and radio access network services for establishing different types of service

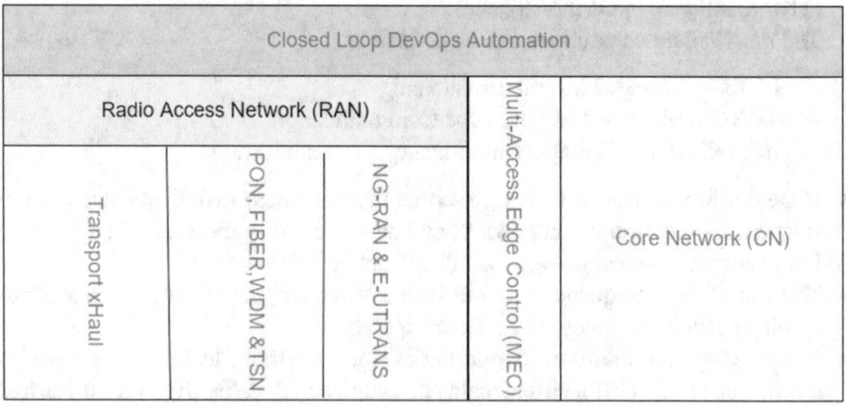

Fig. 5.1 Schematic diagram of different parts of 5G network

controls among the services supported by the most advanced wireless technology in the world.

The radio access network is more advanced that supports numerous additional features and technologies such as eCPRI (enhanced/evolved Common Public Radio Interface), beamforming, MIMO, 5G NR, and many others. The service capability specifications are defined under a range of protocols and standards. Most of the services are developed through software applications running on the modern data networks reducing the dependency on the costly hardware as mostly used in the traditional telecommunication networks [171, 172].

The Closed-loop DevOps Automation is a development and operations management domain in which the repetitive tasks related to operations and maintenance, performance monitoring and correction, addition of new features and configurations to the services and continual monitoring of the 5G services are done in this domain of architecture. The automation of the operations and maintenance is one of the most powerful capabilities of 5G technology that make it one of the most automated communication systems powered by the DevOps methodology and protocols.

The 5G architecture is highly flexible and open for different level of service providers and users to use the architecture through customized service solutions, software infrastructure and applications. The process of using the same infrastructures for numerous virtual networks through virtualization is known as 5G network slicing. The slicing controller works like a control panel used in the cloud computing for creating virtual networks based on the shared infrastructure. This is one of the most futuristic capabilities of 5G network that allows different types of business models used for selling and managing services on the shared as well as dedicated virtual networks. 3GPP has also specified 5G network architecture into multi-layers controlled through a management layer known as network slice controller or orchestrator as listed below:

- Infrastructure layer
- Network function layer
- Service layer
- Network slice controller (orchestrator).

The functional architecture of 5G technology is based on four major layers. Each layer consists of different components—both software and hardware that run on the 5G network architecture. Those four service layers are listed below [172]:

- Network layer
- Controller layer
- Management & Orchestration layer
- Service layer.

In the functional architecture of 5G network, the network layer is one of the fundamental layers in which a range of hardware applications used in the previous technologies such as firewalls, routers, load-balancers, and many others are implemented through Network Function Virtualization (NFV), which are designed to decouple the software from the hardware equipment previously used in older technologies in telecommunication. The most prominent examples of network layer include virtualization of appliances, network slicing technology, and others.

The control layer supports advanced software-defined control functionalities such as Intra-Slice Control (ISC) and Cross-Slice Control (XSC) functionalities that are used to establish the control systems between the network layer and Multi-access Edge Computing (MEC) control systems, which are defined to establish a relationship between the core network functionalities to bring them at the cloud computing edge, which is nearer to the end-users for providing seamless and latency-free services in a shorter span of the distance from the service nodes and the end-users.

The management and orchestration layer of 5G network is highly futuristic layer, which can integrate numerous customized and standard operations and management protocols, tools, methodologies, and procedures. The most common examples of the management and orchestration functionalities supported in the 5th Generation technology include E2E (End-to-end) service management and orchestration, enterprise network management function, 3GPP network management, virtualization MANO (Management and Orchestration), transport network management, and many other customized automation and software defined management services.

The service layer of 5th Generation wireless operations is oriented towards the definition of different types of services defined through software applications in the cloud and connected to the 5G network for public offerings. The most common functions of service layer of the 5G technology include implementation of decision policies, definition of applications and services, and others.

Based on different concepts such as functional, business models, technical, operational, and infrastructure, the fifth-generation networks are classified to be flexible to meet the most complex technical, business and operational needs of the modern world in different scenarios and requirements.

Top Protocols Used in 5G Networks

The fifth-generation network (5G) is the most advanced communication network that uses a wide range of protocols and is capable of communicating with the existing communication protocols and networks along with a huge potential to establish interface with the customized communication protocols to integrate a range of services in the communication models supported by the fifth-generation communication network. A few very important models and protocols used in the 5G networks are described below.

3GPP

The Third-Generation Partnership Project, precisely referred to as 3GPP, is an organization with main focus on developing standards for the wireless telecommunication systems. Primarily, it was established in 1998 to specify the standards for the 3rd generation wireless network mainly focused on GSM core network, but the scope of the work of this standard body evolved and at present, it is the core standard development body behind the latest wireless technology known as 5th generation technology or 5G.

It consists of numerous organization partners that support the development of technology standards, specifications, and protocols for the advancement of the wireless technologies, network element compatibility, forward and backward compatibility, and other issues that can enhance the performance, quality, and value of the wireless communication services in the future. It has the major partners as listed below [173]:

- Association of Radio Industries and Business (ARIB)
- Alliance for Telecommunication Industry Solutions (ATIS)
- China Communication Standards Association (CCSA)
- European Telecommunication Standard Institute (ETSI)
- Telecommunication Standard Development Society India (TSDSI)
- Telecommunication Technology Association (TTA)
- Telecommunication Technology Committee (TTC).

There are three Technical Specification Groups (TSGs) that deal with different domains of wireless technologies for building specification and technical recommendations as listed below:

- Radio Access Network (RAN TSG)
- Service & System Aspect (SA TSG)
- Core Network & Terminals (CT TSG).

Those TSGs are supported by different working groups that meet on a regular basis for updating the standard work on the specified fields.

New Radio (NR)

New Radio is a 5G wireless technology standard developed by 3GPP in collaboration with its partners. This is a standard that defines a wireless interface, which is capable of performing the functionalities defined under 5G network specifications. The 5G New Radio (NR) is specified under two different standards named as:

- **Non-Standalone NR**—This specification is designed for working compatibility with the existing 4G LTE wireless network to have control systems over the 5G data transfer. This standard was started in 1915 and the specifications were completed by the end of 1917. This specification is released under Release-15 of 3GPP.

- **Standalone NR**—This specification was started as the release 16, which was expected to be available in 2020 but due to COVID-19 pandemic, the release was delayed till 2022. Some works have already been done in this direction. This specification defines both control and data transfers under 5G specification without any control compatibility with the predecessor networks like LTE and others in the system. It will be a pure 5G wireless standard that will provide full power of 5G wireless network with high speed and low latency.

The most common features and specifications of 5G New Radio (NR) are mentioned in the following list [174, 175]:

- Uses two frequency band ranges as specified below:

  - Frequency range 1 (FR1)—410–7125 MHz
  - Frequency range 2 (FR2)—24,250–52,600 MHz.

- Dynamic sharing of frequency bands is possible in the non-standalone mode defined under the Release 15 of 5G NR in which the frequency bands can be shared with the existing 4G networks through 4G network controls and compatible with the 5G network capabilities.
- 5G NR supports 5 different sub-carrier spacing as listed below:

  - 15 kHz carrier spacing defined under FR1 with slot duration of 1 ms
  - 30 kHz carrier spacing defined under FR1 with slot duration of 0.5 ms
  - 60 kHz carrier spacing defined under both FR1 & FR2 with slot duration of 0.25 ms
  - 120 kHz carrier spacing defined under FR2 with slot duration of 0.125 ms
  - 240 kHz carrier spacing defined under FR2 with slot duration of 0.0625 ms.

- This standard was built from scratch to choose the best modulation, access technologies, and waveform to produce high speed wireless standard with very low latency and highly efficient usage of wireless resources.
- It is capable of using a range of frequency bands or spectrums such as 2.5–40 GHz including bands like 3.3–3.8 GHz and also using 4.4–5 GHz extensively.
- It supports optimized access technology such as Optimized Orthogonal Frequency Division Multiplexing (O-OFDM).
- 5G NR support beamforming technology for the user-specific signal transmission to provide the most efficient performance of the available wireless resources.
- Support of Multiple Input Multiple Output (MIMO) was also at the core of 4G technology but it will be more efficient and effective technology supported in the NR 5G radio standard such as Multi-user MIMO in combination with the next generation base station known as gNB in the 5th generation network (which takes the performance of the multiple channels to a new height).
- Supports small cell formation to provide higher data throughput near to the end users.

NextGen Core

NextGen Core is a very important part of 3GPP specification for 5G networks. The NextGen Core plays very crucial role in the core network of the 5G wireless technology to materialize high-speed Internet, virtualized services, and low-latency data transmission to meet the ever-growing demand for the Internet services in the world. The NextGen Core specification defines numerous innovative approaches to the core services based on different modern concepts of technologies such as cloud computing, virtualization, network slicing, and so on.

The most important specifications, features, aspects, and characteristics of NexGen Core protocol are listed below [176]:

- Supports numerous techniques, technologies, and schemes at the core-network service plane to produce high level of flexibility, efficiency, and scalability.
- Defines software defined networking, precisely known as SDN, which replaces many hardware parts of the traditional mobile core network with software applications for faster access and operations of the services.
- NextGen Core support network functions virtualization, precisely referred to as NFV, which allows a large range of core network functions defined through software to virtualize and use in different schemes on the same hardware by configuring as per requirement of the services so that high performance core network of functions can be built.
- NextGen specifications define a very powerful functionality in which a particular hardware or core network element can be sliced or specified for numerous types of services as per requirements of the users. Each slice of the network can be used for providing unique and customized service as needed by the end-user. This functionality of NextGen Core makes it highly flexible and customized network that can be used independently for unique services for required level of performance on the same software and hardware resources.
- The performance of the NextGen Core network is defined by Next Generation Mobile Network Alliance, precisely known as NGMN Alliance.

LTE Advanced Pro

Long Term Evolution, precisely known as LTE is 4th generation wireless standard. A few more standards were defined for the enhancement of the LTE technologies and service capacity such as LTE Advanced and LTE Advanced Pro. LTE Advanced Pro, precisely known as LTE-A Pro, is a pre-5G wireless technology that can be easily integrated with the 5G non-standalone standard specified under Release 15 of wireless cellular technology. LTE-A Pro is defined under two releases of 3GPP known as Release 13 and Release 14.

The most important characteristics, features, specifications, and capabilities of LTE-A Pro standard are mentioned below [177, 178]:

- It is commonly referred to as pre-5G, 4.9G, 4.5G, and 4.5G Pro because it is predecessor of 5G technology and are compatible with the 5G technology release 15.
- Supports up to 3Gbps speed and 32 carrier aggregation.
- Supports sharing of licensed and unlicensed spectrum under the scheme of license assisted access spectrum sharing.
- It incorporates massive MIMO technologies in wireless transmission.
- LTE-A Pro deploys 256 QAM (Quadrature Amplitude Modulation) modulation scheme for signal coding.
- It is compatible with the first version of 5G and can be integrated with the existing 5G networks with separate control on the transmission systems.
- It is very useful for data-intensive and critical applications that require huge data rates and lower latency with faster speed.

EPC Evolution

Evolved Packet Core is an LTE standard that combines the voice and data through packet switching. This standard was defined by 3GPP project under Release 8. Earlier than this release, the voice calls would use the circuit switches and data calls would use the packet switches. This release paves the way for advanced core network based on the packet switching.

Impact of 5G Technology on Cybersecurity

The fifth-generation network is very different in many ways in comparison with the existing mobile cellular networks. The most important difference between traditional mobile cellular networks up to 4G and the 5G network is the replacement of numerous hardware devices with the software functions or applications such as firewall, routers, load-balancers, and many others. Thus, the traditional approach to the cybersecurity in the 5G environment will not work properly. It is required to change the way people should think about the cybersecurity in the new environment of 5G technologies.

The most important aspects of the 5G networks that pose serious challenges for the security personnel in the modern cybersecurity sphere include the following [179, 180]:

- Transformation of hardware-based switching networks to software-based routing network, which creates a large number of routing points from where the hackers can intrude into the 5G systems.
- Moving away from the centralized switching systems to distributed systems of routing, which poses another risk for the security of the network.

- Virtualization of network functions poses a serious threat because in the traditional networks of cellular mobiles, the network functions were handled by specified hardware, which is being replaced by the software-defined functions and virtualization of those functions based on the custom-requirements of the users.
- Moving away from the traditional proprietary operating systems on the hardware equipment to the standard Internet protocols poses another very common risk of cybersecurity because every hacker knows about the standard protocols used in the 5G network.
- The network management platform is a software-based application powered by the standard Internet and IP protocols, which would pose a serious risk for the security personnel to cope with. This means, the entire 5G network is very open to attackers to intrude through different access points with its vulnerabilities.
- A huge number of access points and base-station units pose a serious threat at any locality where a hacker can have direct access to the base-station to intrude into the network.
- 5G supports high-speed and high-volume data transmission, which allows users to increase the number of connected devices in the field of the Internet of Things (IoT). Thus, without any proper protocol for the security of the IoT devices, it is a very critical issue to the security of the network of an extensively large network of connected devices.
- Dynamic Spectrum Sharing (DSS) through network slicing is a significant risk to the security of the 5G network because network slicing functionality of 5G network allows multiple streams of information or data to be transmitted through a shared spectrum. The security mechanism should also be dynamic as the spectrum sharing changes.
- There are millions of logged-in mobile users and billions of unregulated entry points to the 5G network which would expand the threat arena multiple times as compared to the traditional cybersecurity threat area.
- Unavailability of security protocols and standards for the Internet of Things (IoT) opens up an expanded risk domain in 5G environment in the form of man-in-the-middle (MITM) attacks through a huge number of connected IoT devices

Based-on the above-mentioned risk factors, it is very clear that the cybersecurity on the 5G should adopt a completely new approaches to cope with the emerging threat domain. The most important steps to properly handle the cybersecurity of the fifth-generation wireless network should include the following:

- A comprehensive cybersecurity plan and investment strategy to manage the security of the 5G networks effectively.
- Proper training of the operators, sub-contractors, and end-users to use the networks and devices more professionally.
- Using VPN (Virtual Private Network) services for connecting devices will be a very good option but may not be feasible for all users and operators.
- A comprehensive collaboration between the operators and governments for devising proper security mechanism and policies is very important.

- A comprehensive collaboration is required among the IoT device manufacturers to follow strict regulations regarding the security of the devices.
- Adaptation of artificial intelligence (AI), machine learning (ML), and other advanced automation technologies for the management of networks is highly desired to cope with the emerging risks of cybersecurity in 5G environment.

Sample Questions and Answers

Q1. State the main features of 1G technology.

A1. The main features of 1G technology include:

- The first-generation technologies started in 1979 when a commercial communication system based on cellular technology was launched in Tokyo, Japan.
- Later in 1980s, numerous countries launched the same in different names such as AMPS (USA), NMT (Nordic), and others.
- The 1st generation technologies used 800 MHz band for voice and signaling based on frequency division multiple access (FDMA) technology.
- This technology was analog wireless technology; so, highly prone to interference and numerous other disadvantages.
- The first-generation technology was limited to just voice calls with no support for text or any other software-based other applications.
- The initial paging networks are also considered as the part of first-generation wireless networks.

Q2. Is the 6G technology available today?

A2. No. The sixth-generation (6G) technology is under development and research phase. This technology has not materialized yet. It can be characterized as an upcoming futuristic technology.

Q3. Write down at least three salient features or capabilities or characteristics of 5G technology.

A3. Three salient features, capabilities, and characteristics of 5G technology are:

- Basic characteristics include larger bandwidths, lower latency, and higher capacity.
- The 5G technology was launched in 2019 and will continue expanding for many years to come to reach at its full capacity and potential.
- The reduced latency can produce the response time to any visual event supported by the 5G technology is about 250 times faster than human response.

Q4. What are the three major components of 5G architecture as the 3GPP specification describes?

A4. The 3GPP specification describes the architecture of 5G technology into three major components as mentioned below:

- Core transport network
- Radio access network
- Service capabilities.

Q5. Write down at least three most important aspects of 5G networks that can pose serious challenge for the cybersecurity.

A5. The most important aspects of the 5G networks that can pose serious challenges for the security personnel in the modern cybersecurity sphere can be:

- Transformation of hardware-based switching networks to software-based routing network, which creates a large number of routing points from where the hackers can intrude into the 5G systems.
- Moving away from the centralized switching systems to distributed systems of routing, which poses another risk for the security of the network.
- Virtualization of network functions poses a serious threat because in the traditional networks of cellular mobiles, the network functions were handled by specified hardware, which is being replaced by the software-defined functions and virtualization of those functions based on the custom-requirements of the users.

Test Questions

1. What is the fifth generation of technology?
2. How does the fifth generation of technology affect our lives?
3. Explain how cellular networks evolved?
4. What are the salient capabilities of 5G?
5. How is the 5G network architecture designed?
6. What are the three Technical Specification Groups (TSGs) that deal with wireless technologies?
7. How does New Radio (NR) work?
8. How does 5G technology impact cybersecurity?
9. How does the 3GPP specification for 5G networks work?
10. What is MANO?

Chapter 6
Internet of Things (IoT)

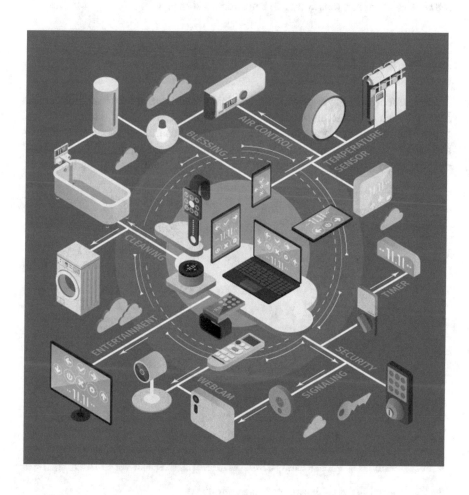

Introduction to Internet of Things (IoT)

Internet of Things, precisely referred to as the IoT, is a relatively new concept of the connected physical devices or appliances through the Internet. In other words, the network of connected devices such as home appliances, office equipment, industrial machines, motor vehicles, and many such types of things that can be controlled through built-in embedded software applications through the intercommunication with mobile applications or other centralized platforms. The connected physical devices have in-built sensors to sense different types of signals such as light, blow, touch, temperature, moisture, and many other things and convert them into digital signals through embedded software applications and get automated instructions to act according to the preconfigured conditions. The Internet of Things (IoT) has expanded exponentially in the recent years. It is expected to cross 22 billion devices by 2025 [181]. The schematic diagram of IoT network is shown in Fig. 6.1.

Importance of IoT

The concept of IoT is so important because it sits at the core of numerous automation schemes such as home automation, office automation, industrial automation, and many other automated processes in all domains of businesses. When all machines and devices can communicate and take action and instructions, the whole world will become very much connected and efficient. The main benefits of IoT include:

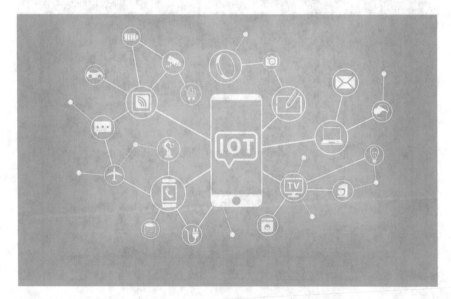

Fig. 6.1 Pictorial concept of Internet of Things. (*flickr*)

- Work efficiency and process productivity
- Effective use of modern technologies for betterment of lives
- Substantial energy and cost saving on different home, office, and business processes
- Reduced footprints of greenhouse gases.

Main Features of Internet of Things

Internet of Things (IoT) is a world of interconnected devices with power to interact and communicate through different sensors and communication protocols. The main features of IoT are mentioned below [182]:

- Effective endpoints management
- High level of scalability, analytic capability, and system integration
- Robust automation and security of the physical assets
- Increased device efficiency and energy saving
- Efficient connectivity among the devices and controlling persons
- Deployment of greater sensing capabilities
- Highly active engagement between the modern technologies and daily-life devices
- Incorporation of cutting-edge technologies like artificial intelligence and machine learning.

History of Internet of Things

The predecessor of the concept of automation through Internet of Things can be traced back in 1950 when the supervisory control and data acquisition (SCADA) system was introduced for the automation of industrial processes based on complex sensing and transmitting systems. The modern concept of automation based on IoT can be marked in 1980, when a code dispenser was modified through computer code by the programmers at Carnegie Melon University, to check the temperature, availability, and waiting queue of coke bottles in the dispenser [183, 184].

- **1999**—The term *"Internet of Things"* was coined by Kevin Ashton and Radio Frequency Identification (RFID) technology was introduced.
- **2000**—The first smart refrigerator was introduced by LG Corporation.
- **2008**—The number of connected devices crossed the number of people on the Earth.
- **2009**—Google Inc started testing the driverless cars.
- **2011**—Google's Nest smart thermostat was introduced in the market.
- **Present**—The IoT world has already expanded tremendously and is expected to make significant impact on all walks of life very soon with the adaptation of 5G technology.

What Is Ambient Intelligence in IoT?

The concept of ambient intelligence, precisely referred to as AmI, is the development of such devices or elements that use pervasive computing, sensing power, adaptivity, and intelligence by collected data from the environment and processing through computing power. The concept of ambient intelligence was coined by Eli Zelkha and colleagues in 1990 [185].

The main examples of ambient intelligence that have used the high-level of sensing devices and artificial intelligence in their invention include Neura app, Otter.ai, ElliQ and others. These services helped a range of categories of people to make their lives so easy.

Autonomous Control in IoT

Autonomous control is a new concept of Internet of Things in which the devices use knowledge-enhanced electronic logic, precisely referred to as KEEL, to make decisions in the connected environment of IoT. As we know, IoT gets the instructions based on a logic when certain conditions are met, but the autonomous control devices use the sensors and artificial intelligence to make decisions for the human being. This concept is known as Internet of autonomous things, precisely known as IoAT. The actuators used in the autonomous things collect and process the information from multiple surrounding sources in the environment and make decisions like the human do [186].

Range of Enabling Technologies Behind Internet of Things

Internet of Things has evolved through numerous concepts that use a range of modern technologies, especially information technology, electromechanical technologies, sensing device electronics, and others. The most common technologies that enabled the concept of IoT to grow exponentially include the following [187]:

- Low Power Sensors
- Cloud Computing
- Artificial Intelligence
- Machine Learning
- Data Analytics
- Big Data
- Enhanced Connectivity
- Internet Protocol V6

Low Power Sensors

The low-power sensors make the IoT devices more useful and efficient for home automation, building automation, and many other automation environments. The notable examples of low-power sensors for IoT include low-power Bluetooth devices, ambient light sensors, smart hubs, digital geomagnetic sensors, digital microphones, programmable push buttons, integrated environment sensors, smart heat sensors, and so on.

Cloud Computing

Cloud computing has revolutionized almost all types of business solutions based on information technology. It offers highly efficient, low-cost, fully-connected, and managed system for building centralized control systems for the IoT systems.

Artificial Intelligence (AI)

Artificial intelligence enables the IoT environments to be dynamic, robust, and inter-active in certain given conditions due to the power of software application that incorporates the artificial intelligence such as similar patterns, figure prints, facial recognition, and many others.

Machine Learning

The capability of machines to learn from the experience or training through data input to consume for understanding the surroundings is another very important technology that enables the Internet of Things, especially in the field of driverless vehicles and aviation domains.

Data Analytics

As the sizes of the interconnected devices increase, the data generated by the devices and systems expand exponentially. To handle that big volume of data, we would need high level of data analysis power. The modern technologies powered by big data and artificial intelligence can handle and process huge piles of data in a very short time period.

Big Data

Big data solutions are very helpful for the IoT environments and solutions because handling of enormously huge data generated by the IoT devices is an uphill task. Big data technologies are assisting IoT systems tremendously to handle the huge volume of data created by those connected devices.

Short Range Wireless Technologies

The short-range wireless technologies are proving to be the backbone of IoT in office, home, building, community, healthcare, and other automation systems. The most important short-range wireless technologies that are highly supportive in IoT environments are [188]:

- ZigBee wireless technology
- Near-Field Communication (NFC)
- Bluetooth and BLE technology
- Radio Frequency Identification (RFID)
- Low-Power WAN (LPWAN)
- Z-Wave Technology
- Wi-Fi Technology.

Medium and Long-Range Wireless Technologies

The medium and long-range wireless technologies that play a pivotal role in the connecting and building centralized control systems in clouds include those technologies that are available in wireless cellular environments. They are used as the backhaul technologies in IoT environments. In both the technical and commercial definitions, the term 'backhaul' generally refers to the side of the network that communicates with the global Internet. A few examples of medium and long-range wireless technologies include:

- 2G/3G/4G technologies
- LTE, LTE-A, LTE-A Pro
- Wi-Max, GSM, CDMA
- Fifth-Generation Wireless.

Effective Communication Protocols

The most important protocols that play very vital roles for the enhancement of IoT solutions and environments are those that operate behind the short, medium, and long-range wireless technologies, sensing communication, control communication and others.

Internet Protocol V6

The number of connected devices has crossed many times the count of the people on the earth. This number will continue to expand exponentially in the future. The IPV4 will not be able to cater to such a huge number of elements in an IoT network. IPV6 is the best option that can help and cater to this huge demand for IP addresses for the connected devices across the globe.

Architecture of Internet of Things Ecosystem

As far as the architecture of the IoT ecosystem is concerned, there is no comprehensive agreement among the researchers on the layers of the architecture of IoT environments. There are different concepts for the architecture of IoT in the marketplace. A few very important architectures are mentioned below [189]:

Three Layer Architecture

Three-layer architecture of IoT consists of three layers that are commonly adopted in the deployment and operations of the IoT networks. They include:

- **Perception layer**—This layer deals with the physical devices, sensors, actuators, and other elements that collect data from the surrounding.
- **Network layer**—The network layer consists of the connectivity of the devices used in the perception layer with the upper layers such as routers, hubs, servers, cloud, and other domains. The processing and transmission of the data is done in this layer.
- **Application layer**—The processing of the sensor related data is done in network layer but in the application layer, the processing of sensor data (for creating an application-specific command and control) is done in the application.

Four Layer Architecture

Another very common architecture of Internet of Things environment is referred to as four-layer architecture as mentioned below:

- **Application layer**—This layer deals with the entire communication or application service between end-device and software application to control and monitor the entire IoT environment.
- **Data processing layer**—This layer handles the sensor data processing and sends it to the application in such a way that it is obtained from reliable/authorized source and is secure or protected from any kinds of threats for the network environment.
- **Network layer**—It is also known as transmission layer. The main responsibility of this layer is to connect the network elements, i.e., sensors, servers, routers, and other intermediatory devices and transmitting data between those devices and application layer.
- **Perception layer**—This is like a physical layer that consists of different elements or end-devices in an IoT environment. It is also known as sensor layer, which generates data based on the environmental conditions and processes that to digital information and sends to the network layer.

Five Layer Architecture

The five-layer architecture of IoT environment is more comprehensive with many functionalities and processes. The five layers of this architecture are:

- **Perception layer**—This like a physical layer and deals with the sensors, actuators, sensor embedded devices physically. This layer is similar to that of the three-layer model.
- **Transport Layer**—The transport layer deals with the communication to and from perception layer to the processing layer of this model. It deals with the interconnectivity of end-devices and the network elements such as routers, hubs, servers, and others.
- **Processing layer**—This layer deals with the data storing, processing, and analyzing of huge piles of data created through end-devices. The results are transported to the transport layer for further communication. It is also known as middleware layer of IoT architecture.
- **Business layer**—This layer deals with different business models used in the IoT business environment. The cloud-based different models are deployed under this layer of IoT ecosystem.
- **Application layer**—This is the upper layer of this model, which deals with the application services to provide the application-specific information to the end users of the IoT environment. The interface between the end-user and IoT system is established through this layer for monitoring and operational purposes.

What Is Decentralized Internet of Things Concept?

Decentralized IoT is a modern concept motivated by the prospects of highly cutting-edge technologies such as distributed ledger blockchain technology, multi-access edge computing (MEC), artificial intelligence (AI), and machine learning (ML). These technologies implemented into the IoT ecosystem provide highly effective, cost-efficient, reliable, secure, and robust solutions to the businesses. The deployment of MEC and blockchain technology into IoT is known as smart IoT or decentralized IoT. MEC improves scalability, efficiency, performance, and software defined resources to the nearest point in the cloud and blockchain develops high level of trust, security and reliability to the systems [190].

What Is Industrial Internet of Things?

Industrial Internet of Things, precisely referred to as IIoT, is also known as Industry 4.0 in the modern world. The IIoT is the concept of interconnection of a wide range of industrial sensors, actuators, machines, processes, and human beings simultaneously to enhance the level of automation in the industrial processes. This revolutionizes the industrial automation due to adaptation of power of artificial intelligence, machine leaning, big data analytics, cloud computing, edge computing, and many other modern technologies. The use of robotics with the control applications and other actors of different industrial processes are governed by numerous enabling technologies to take industrial automation to a new height.

Industrial Internet of Things Standard Bodies

All international and regional standard bodies that define different types of standards for different technologies and platforms are relevant to IIoT. In addition to those standards bodies, a few more domain-specific bodies that play very vital role in the enhancement of Industrial Internet of Things include the following:

- **Industrial Internet of Things Consortium (IIC)**—This is an object management body to provide standards useful for enhancing transformative business value in the industries. It was established in 2014 and provides the fundamental support on enhancing transformation of digital processes in a range of industries.
- **IoT Acceleration Consortium**—The main objective of this standard body is to combine different players such as academia, governments, industries, and research organizations to build a structure for developing, deploying, and operating the best technologies for industrial automation to enhance revolution 4.0 in the industries.
- **Alliance for Internet of Things Innovation (AIOTI)**—This is a very strong alliance of European academia, industries, IoT players, SMEs (Small and

Medium-sized Enterprises), and others with a focus on the enhancement of creating dynamic European IoT ecosystem. This alliance was established in 2016.

- **Open Platforms Communication Foundation (OPC)**—The main objective of this platform is to build standards for the software developers, end-users, application platforms, industries, and other players for the smooth operations with strong interoperability of the devices, applications, interfaces, and operations and maintenance.

As mentioned earlier, IoT standards and regulations are under development and many of them have not yet matured, especially in the field of security, privacy, network connectivity policies and many other operations related matters. The above-mentioned bodies along with numerous other traditional international standard bodies are continuously working to streamline the development, deployment, operations, security, and privacy policies of this fast-growing domain of information technology.

Important Industrial Internet of Things IIoT Platforms

Industrial automation has become an integral part of almost every industry, especially mining, IT, oil and gas, aviation, automobiles, manufacturing, processing, operations and maintenance, community management, smart cities, civic utilities, etc. According to the Valuates Research information, the global market size of IIoT platforms is expected to cross USD $102.4 billion by 2028 from USD $71.16 billion in 2021 with a graceful growth rate of over 5.3% CAGR over the forecast period between 2022 and 2028 [192]. There are numerous major players with a sizeable share in the global IIoT platform market. A few of them are listed below:

- Azure IoT
- Oracle IoT Cloud
- IBM Watson IoT
- AWS IoT
- Siemens Mind Sphere
- Flutura Cerebra
- Thing Worx
- GE Predix.

Let us have a look into them individually with their respective features, capabilities, characteristics, and applications in certain industries.

Azure IoT

Azure IoT platform is a professional-grade IIoT application that supports numerous features such as SDK (Software Development Kit) for development, cloud computing, faster deployment, effective contracting and evaluation, and best support services. This platform is offered by Microsoft Corporation, which is an American technology giant.

Oracle IoT Cloud

Oracle IoT platform is a cloud-based efficient application for managing a wide range of devices under a unified monitoring and maintenance. It supports range of capabilities, especially big data management, data analytics, KPI (Key Performance Indicator) settings, performance monitoring automated functions to take action against any kinds of operational anomaly or security threats.

IBM Watson IoT

IBM's Watson is a highly powerful IoT system that uses high level of cognitive intelligence in its functionalities. It can support automation environment powered by machine learning, artificial intelligence and data analytics. It can support capabilities of large-scale device integration, operations and maintenance, data analysis, data virtualization, etc.

AWS IoT

Amazon Web Services (AWS) is the largest player in the cloud computing solutions. It offers professional platform for managing industrial devices in IoT environment. It is a very comprehensive platform that supports a wide range of operations and management environment that are based on high-grade automated solutions.

Siemens Mind Sphere

Siemens is one of the leading manufacturing companies based in Germany. The Siemens PLC automated systems for manufacturing industries are very well known worldwide. The MindSphere is the latest industrial automation platform from Siemens for effective O&M (Operations and Maintenance), integration, and security solutions in a range of manufacturing processes.

Flutura Cerebra

Cerebra IIoT platform is offered by Flutura. This is one of the pioneers in the IIoT software management and automation. It supports numerous modules and IoT networks for unified automation solutions in the industries. It is easy to add, configure, deploy, and manage a range of modules in IIoT ecosystem.

Thing Worx

Thing Worx is a comprehensive and end-to-end industrial automation platform that can integrate a variety of devices in IIoT ecosystem and collaborate through secure and reliable communication systems in such a way that an industrial-grade automation system is achieved for the operations and maintenance of a range of industrial processes.

GE Predix

General Electric Inc (GE) offers a cloud-based digital application for IIoT automation ecosystems. It is highly flexible, secure, scalable, and reliable platform that offers edge-to-cloud automation services in the IoT environment. It is a modular system that can support a range of modules such as HMI (Human Machine Interface), SCADA (Supervisory Control and Data Acquisition), ERP (Enterprise Resource Planning), Controller, Sensor, and others.

IIoT Use Cases in Different Industries

Industrial Internet of Things (IIoT) can be used in numerous applications across the industries. Many new concepts have been created due to the emergence of modern IIoT in different industries. The connected world of devices based on IIoT is used in many domains as example use cases. A few of them are listed below:

- Smart Cities
- Smart Home
- Manufacturing
- Process Automation
- Energy Management
- Supply Chain
- Healthcare
- Agriculture

- Military
- Transportation.

Let us know more about the use cases of Industrial Internet of Things in different domains of industries and walks of modern community management systems.

Smart Cities

The use of industrial IoT in smart cities is taking the center stage in the management of traffic, water supply, electricity, drainage systems, and many other civic services by using IIoT connectivity. Smart cities use numerous types of light sensors, water sensors, imaging sensors and other devices for collecting data from numerous systems in the modern cities and devise a comprehensive solution for managing the systems effectively and efficiently.

Smart Home

Smart home is another use case of IoT. A smart home is that one in which the lights, heaters, air-conditioning equipment, physical security, access control, cooking, and other processes are automated through integrated ecosystem based on IIoT platforms. The sensors send data to the controlling unit for processing and taking the suitable automated decisions based on pre-defined criteria of operations.

Manufacturing

Manufacturing has been using automated systems based on PLC (Programmable Logic Controller) for many years now. To clarify, a PLC is an industrial computer control system that continuously monitors the state of input devices and makes decisions based on a custom program to control the state of output devices. With the advent of modern IIoT platforms, the automation in manufacturing based on IoT platforms has taken a central position instead. A network sensor could be deployed to collect manufacturing process data. Again, it can be used to assess the efficiency and performance of the processes in manufacturing. The customized manufacturing, robotics, and safety mechanism are a few processes to name for this.

Process Automation

The automation of numerous processes that involve data from different elements to process and respond accordingly has been done through the usage of IIoT. Many operations and maintenance processes, data analysis processes, reporting and others are extensively automated through modern IIoT platforms.

Energy Management

The analysis of energy usage and its patterns in different conditions and circumstances are analyzed and proper plan are devised on the basis of that useful insight of the energy consumption in any industrial or commercial activities. This is known as energy management through automated system of connected devices. An IIoT-based solution for energy management can integrate multiple sources of energy and manage the most efficient use of the energy in your businesses.

Supply Chain

Supply chain has become highly automated in recent years as compared to the traditional models of supply chain management. Supply chain management is using numerous processes from quality control through delivery of the products to the client that are controlled by the automated processes managed by the IoT and associated software platforms. With the passage of time, the supply chain management will become fully automated and mechanized under the control of IIoT models.

Healthcare

Many automatic health monitoring tools have been introduced in the market place that can be governed and controlled through IIoT platform under the centralized control systems. The tracking of patient health, medical records, and doctor advises are possible to be monitored through IIOT platforms. The automated management of healthcare processes for elderlies are also extensively monitored through IoT network effectively.

Agriculture

Automation of irrigation systems based on the field monitoring and analysis results is known as automated agriculture. Modern IoT-based field monitoring and analysis systems have been developed to track different parameters such as crop density, moisture, temperature, humidity, crop health and other factors. Based on those results, an automated solution is devised to provide the required support to the crop in the form of watering, soil manuring, pesticide spraying, and other processes.

Military

An integrated and comprehensive ecosystem based on IoT is being used by almost all militaries in the world to track the positions of their own resources and spy on the resources and movements of the adversaries. A unified network of automation of the information collected from a range of sensors and radars can help the militaries extensively.

Transportation

Transportation is using integrated network of different elements of transportation networks such as tolls, roads, vehicles, speeds, traffic, and associated components. The major applications of IoT in transportation industry include, automated traffic management and signaling systems, toll collection, vehicle tracking, vehicle tax collection and monitoring, public transportation management, route management and other activities.

Challenges Posed by Internet of Things

The field of IoT is expanding unprecedentedly with millions of new devices getting added in just the matter of days and weeks. In such huge growth and unavailability of comprehensive regulation and control protocols, IoT poses numerous challenges for all concerned people in this field. Those challenges can be classified into many categories. A few of those main challenges associated with IoT are listed below:

- Cybersecurity
- Privacy
- Complex Operations and Management
- Environment Impact
- Bulky Data.

Let us have a closer look at those challenges to understand the reasons behind them.

Cybersecurity

Cybersecurity is one of the most important challenges posed by the ever-expanding environment of IoT worldwide due to numerous reasons and conditions. This topic will be dealt separately in the closing section of this chapter.

Privacy

When every device, equipment, appliance, or item that people use is exposed to the public through different ways such as monitoring and tracking systems, through less secure network, and multiple points of intrusion, the privacy of the users will be on stake for sure. The main reasons and causes of the violation of privacy of a person are directly associated with the security of the networks, which will further be discussed in the closing part of this chapter.

Complex Operations and Management

The operations and management devices are manufactured by a huge number of different manufacturers who use proprietary platforms and operating systems to run their products. This creates a great complexity to integrate all those types of devices, equipment, and items into an automated monitoring system that can support a range of operating systems, device features, specifications, and capabilities. Thus, the configuration, monitoring, and maintenance of the operations of the diverse set of equipment is a big challenge in the IoT environments. In fact, the absence of end-to-end compatible protocols, regulations, and agreement on procedures, makes the challenge even more difficult to cope with.

Environment Impact

Continuously increasing demand for the power and batteries used in the equipment will impact critically on the environment one way or the other. For instance, additional batteries for the equipment, routers, control systems, local equipment will require additional landfills/dumping yards and the impact on the environment will also be somewhat adverse.

Bulky Data

The IoT-based networks expand the volume of data rapidly. This data created by such gigantic networks will become so bulky that it will pose a big challenge to deal with or to deduce any valuable information from that data. The concept of big data basically deals with this and researchers are trying to come up with effective solutions as the data would be produced in terabytes. According to an estimated projection (at the time of writing this book), about 463 exabytes (1 exabyte = one billion gigabytes) of data will be created on a daily basis in 2025 [193]. Managing such a huge pile of data will become a big challenge for the concerned industry experts and technologists.

Impact of IoT on Cybersecurity

The impact of IoT environment on cybersecurity is huge due to many reasons. To understand those main causes associated with the IoT ecosystems that impact the cybersecurity can easily be understood when the main features of IoT are figured out (i.e., features that are directly related to cybersecurity). Let us figure out the key features of IoT ecosystem that would directly impact the cybersecurity [194].

- IoT is expanding exponentially without any strict rules and regulations
- IoT ecosystem is not mature yet and it will take time to get real maturity
- Number of devices is enormous
- All devices are not so powerful to support security measures and ensuring that is quite difficult
- Professional management of such a large network of devices is not possible always
- Compatibility and interoperability are other major issues
- Low power devices sometimes cannot connect with other devices continuously
- Diverse and easy to access locations for hackers is a big challenge for security
- Large volumes of data are stored in the cloud from where the data can be stolen
- Regular monitoring and upgrading of the devices are some big challenges
- Password management is very difficult in a large number of devices
- Lack of trained personnel and users with awareness about cybersecurity is a major challenge
- Lack of proper legislation and regulatory laws (which would work anywhere as it demands)
- Unavailability of in-built security features and capabilities in all types of devices used in IoT networks.

All above-mentioned characteristics of IoT networks or ecosystems pose serious threats to cybersecurity. The hackers can get physical access to devices connected to the IoT network. The number of devices increases the chance of hackers to intrude into the system. The less in-built security and devices with low computing power provide an easy way to intrude into the systems, especially into the cloud to compromise the valuable data. Managing and updating all software and firmware programs is not possible for such a huge number of devices because there is not enough sufficiently skilled manpower to carry out such a huge activity on a regular basis. This problem may not ever be solved! In such circumstances, the challenge to the cybersecurity is huge in the IoT ecosystems.

Sample Questions and Answers

Q1. What is ambient intelligence?

A1. The concept of ambient intelligence, precisely referred to as AmI, is the development of such devices or elements that use pervasive computing, sensing power, adaptivity, and intelligence by collected data from the environment and processing through computing power. The concept of ambient intelligence was coined by Eli Zelkha and colleagues in 1990.

Q2. What are the three layers in the three-layer architecture of IoT ecosystem? Briefly describe each of them.

A2. The three layers are:

- **Perception layer**—This layer deals with the physical devices, sensors, actuators, and other elements that collect data from the surrounding.
- **Network layer**—The network layer consists of the connectivity of the devices used in the perception layer with the upper layers such as routers, hubs, servers, cloud, and other domains. The processing and transmission of the data is done in this layer.
- **Application layer**—The processing of the sensor related data is done in network layer but in the application layer, the processing of sensor data (for creating an application-specific command and control) is done in the application.

Q3. What do you mean by IIoT?

A3. IIoT stands for "Industrial Internet of Things". It is also known as Industry 4.0 in the modern world. It is the concept of interconnection of a wide range of industrial sensors, actuators, machines, processes, and human beings simultaneously to enhance the level of automation in the industrial processes. This revolutionizes the industrial automation due to adaptation of power of artificial intelligence, machine leaning, big data analytics, cloud computing, edge computing, and many other modern technologies. The use of robotics with the control applications and other actors of different industrial processes are governed by numerous enabling technologies to take industrial automation to a new height.

Q4. What is Azure IoT?

A4. Azure IoT platform is a professional-grade IIoT application that supports numerous features such as SDK (Software Development Kit) for development, cloud computing, faster deployment, effective contracting and evaluation, and best support services. This platform is offered by Microsoft Corporation.

Q5. What are the challenges posed by IoT environment for operations and management tasks?

A5. The operations and management devices are manufactured by a huge number of different manufacturers who use proprietary platforms and operating systems to run their products. This creates a great complexity to integrate all those types of devices, equipment, and items into an automated monitoring system that can support a range of operating systems, device features, specifications, and capabilities. Thus, the configuration, monitoring, and maintenance of the operations of the diverse set of equipment is a big challenge in the IoT environments. In fact, the absence of end-to-end compatible protocols, regulations, and agreement on procedures, makes the challenge even more difficult to cope with.

Test Questions

1. How does the Internet of Things (IoT) work?
2. How does Ambient Intelligence work on the IoT?
3. What are the enabling technologies behind the Internet of Things?
4. How do Short Range Wireless Technologies work?
5. How does Internet Protocol V6 work?
6. How does the IoT ecosystem work?
7. What Is the Concept of Decentralized Internet of Things?
8. Why should I use AWS IoT?
9. In what ways does the Internet of Things pose challenges?
10. IoT and cybersecurity: what are the impacts?

Chapter 7
Distributed Cloud Computing

An Introduction to Distributed Cloud Computing

The concept of distributed cloud computing is a part of cloud computing systems in which the computing resources and software applications are located at different cloud service providers or servers but they act under one cloud computing software for operating and managing the resources across multiple cloud platforms. The

K. Thakur et al., *Emerging ICT Technologies and Cybersecurity*,
https://doi.org/10.1007/978-3-031-27765-8_7

distributed cloud may consist of numerous types of cloud services and may be offered by different cloud providers under one centralized management system that acts as a one management plane for all those services located in different geographical locations under different service providers. In other words, it is an architecture of multiple cloud computing types, services, and providers that are located at different locations and data centers and they operate like a single integrated and centrally managed service.

The main features and characteristics of distributed cloud computing systems used in the modern cloud environment include [195, 196]:

- A structure of multiple software components that reside at different locations, under different service providers for providing multiple types of cloud services under one single centrally managed cloud service is known as distributed cloud computing service.
- Those isolated services at different locations can be either connected through local networks or wide area networks for remote services.
- This service many include different types of cloud services such as infrastructure, platforms, software, data storage, and any other third-party service.
- The entire service may include a range of machines such as mainframes, personal computers, minicomputers, workstations, and other devices.
- The entire service works like a single unified and centrally-controlled mechanism through a single service management and operations platform.
- Distributed cloud system is completely independent of the underlying software applications, types of infrastructure, geographical location, and types of services offered by any third-party service provider.
- Distributed platform can integrate different types of operating systems and communication protocol stacks such as Linux, Unix, Windows, Ethernet, Token Ring, TCP/IP (Transmission Control Protocol/Internet Protocol), or any other network protocol.
- It is a client/server architecture or model that operates independently and seamlessly. It is also known as three-tier client–server model.
- The three-tiers of distributed cloud computing model include Presentation (Tier-I), Business logic (Tier-II), and Data/Resources (Tier-III).
- Usually, the distributed cloud service is offered over public cloud that integrates multiple public cloud, private clouds, on-premises infrastructure or software, and third-party cloud services under one roof.
- It supports multi-cloud and hybrid cloud environments simultaneously. In other words, it is a unified cloud service that runs numerous types of cloud computing services as the basic components operating under one roof without any discontinuity or disconnection known to the end-users of the services.
- It is a complementary cloud computing service to edge cloud computing concept which will be covered next in this chapter.

What Is Edge Computing?

The concept of edge cloud computing is somewhat similar to the distributed cloud computing environment. In edge computing, the real-time data or content service is provided to the end-users at the edge of the environment very close to the end-users. For example, content delivery network (CDN) is one of the examples of an edge computing in which the content (images, video) is provided through multiple servers or service providers at different locations very near to the end-users that seek that content across different geographical locations (in the globe). The main features and characteristics of edge cloud computing include [197]:

- The collection, processing, storage, and provisioning of the data generated in the modern high-speed and real-time services powered by IoT and 5G technologies is normally required to be carried out over the edge of the cloud computing not at the data centers or the service centers located in the cloud at remote cloud service point. The collection, processing, storage, and provisioning of data near to the end-user is known as cloud computing services, which can be any type of service located at the edge of the cloud for faster delivery of processing, storing, and provisioning capability.
- The application of edge-computing is increasing due to the increasing demand for real-time data processing and utilization in real-time data-driven decision-making and real-time service processing applications powered by artificial intelligence (AI), machine learning (ML), IoT, and other similar kinds of advanced technologies.
- The core objective of edge-computing is to provide the real-time data processing and provisioning to all stakeholders of a service network seamlessly irrespective of their location and roles in the networks.
- The complete data-flow of an organization is made equally available to the entire network user for efficient and effective use of the data/information in decision making.
- The data is mostly generated at the edge of the cloud in different devices such as IoT devices, sensors, computers, end-users, and so on. The processing of that data at the central location will make the network very slow, highly latent and inefficient for real-time decision making of the actions based on artificial intelligence. To cope with this bottleneck, the edge computing comes into play to process the data at the edge of the cloud and provide it to the stakeholders in real-time and in an efficient manner. At the same time, the data is stored at local locations as well as the central locations for easy and instant access.

Advantages of Distributed Cloud

Distributed cloud computing is an advanced version of cloud computing to offer numerous additional features and benefits to the businesses as well as to the end-users in modern era of communication powered by fifth-generation wireless communication and exploding field of the Internet of Things (IoT). The most important benefits of distributed cloud include [198]:

- **Regulatory compliances**—The regulators from different countries, regions, and service-domains have set different regulations and compliance standards to meet in terms of data privacy, security, service quality, and many other factors. The distributed cloud computing offers a very flexible and effective solution to meet the regulatory compliances in all regions and countries effectively and easily.
- **Flexibility**—The distributed cloud offers a high-level of flexibility in deployment, installation, relocation, debugging, provisioning, and security management of services, data, and network environments.
- **Scalability**—Distributed cloud computing systems offer better scalability for expanding different nodes, virtual machines, and other cloud resources and nodes in very short time without any major overhauling of services and infrastructure.
- **Enhanced performance**—The distributed computing offers much better performance and speed as compared to the traditional cloud computing systems. This offers the least network latency and services near to the end-users.
- **Faster processing**—The distributed cloud computing offers faster processing power due to the power of processing located at multiple locations and integration of large number of cloud resources across multiple locations and services of the network.
- **Greater uptime**—Distributed cloud computing offers increased uptime of service due to multiple servers, clouds, and services operating simultaneously from multiple locations.
- **Security**—Distributed cloud networks offer greater level of security and control due to the availability of options for multi-cloud and multi-vendor support.

Working Principle of Distributed Cloud

The basic working principle of distributed cloud computing is to distribute or split one single task among a number of virtual machines located at different locations to provide faster processing, greater efficiency, and reduced latency in the network. The virtual machines handling the tasks are networked together for effective collaboration and coordination for producing the desired level of performance, security, quality, and latency [199].

Distributed Cloud Architecture

The architecture of distributed cloud can be divided into multiple categories. The most fundamental categories of cloud computing architecture are [200]:

- System architecture
- Software architecture.

The system architecture can further be divided into two categories based on their mode of communication in a network environment.

- Client–Server Architecture
- Peer-to-Peer Architecture.

The schematic diagram based on the definition of a distributed cloud computing system is the combination of multiple types of computing resources that work under a unified cloud management architecture to distribute and process the data in separate virtual machines and other nodes such as workstations, on-premises data service, and other computers working as third-party service processing elements. The schematic diagram of distributed cloud computing system consisting of different elements and nodes is shown in Fig. 7.1.

The distributed cloud computing system architecture can be further divided into four classes based on the software systems as listed below:

- Layered architecture
- Data-centered architecture
- Event-based architecture
- Object-based architecture.

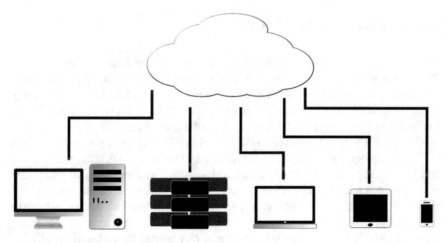

Fig. 7.1 Schematic diagram of distributed cloud. (*pixabay*)

All these software-based architectures work in different ways to meet the specific requirements of a particular model or architecture.

Top Use Cases of Distributed Cloud in Industries

The use of distributed cloud computing is increasing very fast across the industries worldwide. The global market size of distributed cloud was just about USD $1.3 billion in 2020, which is expected to reach USD $5.0 billion by 2026 with a whopping growth of over 26.5% CAGR over the forecast period between 2021 and 2026 [201]. This huge growth in the market size of distributed cloud is driven by a wide range of use cases adopted in the industries. A few of the most important use cases popular in the industries are mentioned in the following list:

- Content Delivery Network (CDN)
- Internet of Things (IoT) and Edge
- Software Defined Infrastructure (SDI)
- Big Data Processing
- Multi-Cloud Unification
- Centralized Management.

The number of use cases of distributed cloud is continuously increasing. Many applications powered by the modern technologies such as artificial intelligence, machine learning, big data, data analytics, cognitive learning, 5G technologies, IoT, etc. can use this. Let us now discuss these technologies with a closer look.

Content Delivery Network (CDN)

CDN stands for "*Content Delivery Network*" or "*Content Distribution Network*". This is a relatively new concept of distribution of heavy content that usually consumes substantial volumes of bandwidth and time to load to the websites requested by any user across the globe. CDN handles the heavy content in such a way that it improves the loading time of the websites and enhances the user experience on the websites.

Content delivery network (CDN) is a network of proxy servers that are situated at different geographical locations in such a way that they provide the content very fast by reducing the distance of content to be loaded from the main server located at very remote location. The proxy servers cache the content of the main server and provide that content on request of the users in such a way that the nearest server will respond to the request for content on the network. There is a network of proxy servers in which the proxy servers are located very near to the end-users usually.

The main points where the CDN proxy servers are located are called the Internet Exchange Points (IXP), which are the locations where different networks or communication systems exchange information among separate networks. For example,

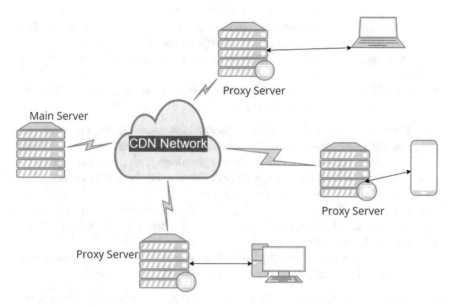

Fig. 7.2 Schematic diagram of CDN network

London could be the best location for placing CDN servers for the users that use different networks, which exchange their inter-networks information through many major telecommunication players like British Telecom (BT) and other European countries. Those servers are very useful for providing faster delivery of different types of content in the United Kingdom and its surrounding countries. If the demand for the data increases, the CDN servers can be added in each country or even in the major cities of a country to cater the increased demand for the content. The content provided by the CDN servers is in static forms such as images, videos, documents, client-side scripts, and others. A schematic diagram of a CDN network is shown in Fig. 7.2.

The main features, capabilities, and characteristics of content delivery network are as follows [202, 203]

- It is a distributed network of data center and proxy servers to provide heavy content as fast as possible for providing webpage loading efficiency and performance.
- The history of CDN can be traced back to 1990 to enhance the performance of Internet in those days, but the market expanded after the advent of modern technologies such as IoT, big data, and others.
- The main business activities that are influenced by the CDN include software downloads, license management, video streaming, website loading, transparent caching, real-time data analytics, load-balancing, and others.
- The routing to the right proxy server in CDN is done through automatic process based on a comprehensive algorithm(s). For this purpose, numerous protocols are

used such as Internet Content Adaptation Protocol (ICAP), Open Pluggable Edge Service (OPES), Edge Side Includes (ESI), and others.

- There are two major types of CDN networks such as Peer-to-Peer CDN and Private CDN.
- The working procedure of CDN networks is based on layered-functionalities, which include the following layers:

  - **User layer**—This includes the visitors and management of the network.
  - **Network switch**—Establishes the connectivity based on hardware address.
  - **Routing layer**—Deals with the software address layer of the connectivity.
  - **Catching servers**—This layer includes the network of catching servers located at multiple locations near to the end-users.
  - **Scrubbing servers**—A layer of dedicated machines that receive all network traffic destined for an IP address (or IP addresses) and they attempt to filter good traffic from bad.
  - **Original server**—This is the lowest level where the main CDN server resides for the delivery and storage of the main content.

- The main objectives of using CDN network is to reduce the latency of the communication network, increase the performance and efficiency of content-centric applications, reducing the loading period of websites, increasing the security of the content, and many more.

Internet of Things (IoT) and Edge

The IoT and edge computing are other major usage scenarios that require the power of distributed cloud computing technology. The details of IoT and edge cloud computing have already been provided in the previous chapter. The readers can refer to the previous discussions and we prefer not to repeat the same information in this chapter.

Software Defined Infrastructure (SDI)

Software defined infrastructure is another very emerging field that can use the power of distributed cloud computing more effectively and efficiently. The control of hardware-based computing resources such as storage, memory, processors, networking, and allied equipment through software platform is referred to as software defined infrastructure. The concept of cloud computing through virtualization of computing resources is an example of software defined infrastructure. In other words, the automated control of the hardware through software program can be termed as software defined infrastructure or precisely, SDI [204].

The automation of a wide range of business processes powered by the software-based infrastructure without a substantial intervention of human is known as software defined infrastructure. It is still a bit vague definition in which the level of automation is not expressed explicitly. In today's world, there are numerous software-based platforms that define, configure, and control computing resources virtually with the help of software platforms. A few examples of software defined services are mentioned below:

- Software-defined compute (SDC)
- Software-defined storage (SDS)
- Software-defined data center (SDDC).

The SDI allows the organizations more flexibility and automation so that the manpower can focus on other core issues rather than maintaining routine processes.

Big Data Processing

Big data processing is a very important use case of distributed and edge computing in the modern business data management. The most important sources for generating the big data include machine-generated data, especially in IoT environment, digital transactional data, and social media platforms. The involvement of artificial intelligence in the edge computing, which is a part of distributed computing is extensive in processing the big data at the edges of the data sources. This is used in such a way that the real-time business information can be extracted and acted upon automatically to create a value for the businesses. All major components of big data such as data collection, data tagging, data processing, and data distribution and storage are all very complex and bulky processes that need huge computing resources, which are arranged and managed through distributed cloud computing and edge cloud computing [205].

Multi-cloud Unification

Unification of multi-cloud with hybrid, edge, and distributed cloud enhances the capabilities and automation level of an organization tremendously. The unification of multi-clouds is designed to use the power of distributed computing system, especially the edge computing to process the data from the major data sources efficiently and quickly. The distributed cloud processing, through different clouds such as hybrid, on-premises, edge server cloud, and other make a great use case scenario for all types of process automations across the industries [206].

Centralized Management

Centralized management of company resources, processes, and stakeholders is referred to as centralized management in modern field of IT terminology. The centralized management became possible and more effective due to the utilization of distributed and edge cloud computing. The management of all major processes based on the distributed cloud computing in remote location became possible to be managed through centralized management platforms that support distributed technology automation.

Challenges of Distributed Cloud Computing

The distributed cloud computing technology offers a range of benefits and desirable technical capabilities that make the process automation of modern business more efficient, beneficial, flexible, scalable, and reliable. Despite the benefits and desirable aspects of distributed cloud computing, there are certain challenges that make the decision makers hold back from adapting this technology at full scale. Researchers are trying to come up with effective solutions to tackle those challenges. A few very important challenges associated with the distributed cloud computing are mentioned below [207].

- **Complexity**—The distributed cloud becomes more complex due to its nature of heterogeneity of systems, clouds, and computing resources. Managing a complex architecture may pose a big challenge for many technical teams.
- **Security**—The security of distributed cloud systems is a very big challenge, which leaves direct impact on the overall cybersecurity mechanism and strategy of an organization. The impact of distributed computing on cybersecurity will be discussed as the next topic.
- **Contingency**—The contingency strategy to cope with any kinds of mishap or attack is a complex matter to deal with. Backing up and restoring a distributed network from different locations of the world is a very difficult and time consuming task.
- **Bandwidth**—If the distributed components communicate with the centralized computing resources in a certain datacenter, the bandwidth will become fully congested because different nodes at different locations in distributed system would communicate with the centralized datacenter. The edge computing may be deployed to overcome this problem.
- **Transparency**—Transparency of different separated elements sometimes conceals things that may lead to some serious issues related to security, policy, and strategies in the technical operational environments.
- **Technical faults**—The number of technical faults at both software and hardware levels are likely to happen much more than the traditional centralized cloud computing or on-premises solutions because the number of interfaces used in

distributed cloud are very high. Those interfaces may become point of faults or failures in certain conditions.

- **Troubleshooting**—The troubleshooting of a large and complex system is always difficult; so is the case of distributed cloud architecture. It is indeed difficult to debug errors in software code or fix a fault in the complex network based on multiple hardware and software components located at different locations and on virtual machines.

Impact of Distributed Cloud Computing on Cybersecurity

The impact of distributed cloud computing on cybersecurity is very high due to the emerging new solutions based on distributed cloud in which a scattered network of edge servers and processing computers are located at different locations to process the data very near to the data generating sources. The expansion of IoT network is also a big challenge for the cybersecurity in the distributed cloud environment. The most important impact of those cybersecurity challenges result in substantial business losses to the businesses and organizations [208].

The mentionable impacts left by the distributed cloud computing on cybersecurity include the following:

- **Extended attack surface**—The attack surface is defined as the number of all possible points, or attack vectors, where an unauthorized user can access a system and extract data. The attack surface has expanded significantly due to the extensive use of distributed cloud computing services in public cloud domain. A huge network of cloud components with multiple interfaces, vulnerabilities, and bugs opens up a new attack surface for the hackers. The hackers have already become technically so advanced that they adopt numerous techniques to exploit the extended attack surface.
- **New threats**—Hackers are using new types of threats more advanced than the traditional kinds of threats such as payload downloaders, remote admin toolkit (RAT), service boosters, and other similar kinds of advanced threats.
- **Complex network**—The distributed cloud has made the enterprise networks more complex, large, and difficult to manage. A big network, especially the distributed cloud network with IoT devices, is very difficult to manage due to numerous reasons such as lack of comprehensive regulation, rules, standards, protocols and security strategy.
- **Security policy**—The traditional security policies are not sufficient to cope with the extended threat landscape in distributed cloud network. A new security policy with proper automated processes (to handle a huge network) should be adopted; otherwise, the networks would remain at risk all the time.

Sample Questions and Answers

Q1. Define Distributed Cloud Computing.

A1. The concept of distributed cloud computing is a part of cloud computing systems in which the computing resources and software applications are located at different cloud service providers or servers but they act under one cloud computing software for operating and managing the resources across multiple cloud platforms. The distributed cloud may consist of numerous types of cloud services and may be offered by different cloud providers under one centralized management system that acts as a one management plane for all those services located in different geographical locations under different service providers. In other words, it is an architecture of multiple cloud computing types, services, and providers that are located at different locations and data centers and they operate like a single integrated and centrally managed service.

Q2. What is Edge Computing?

A2. The concept of edge cloud computing or simply, edge computing is somewhat similar to the distributed cloud computing environment. In edge computing, the real-time data or content service is provided to the end-users at the edge of the environment very close to the end-users. For example, content delivery network (CDN) is one of the examples of an edge computing in which the content (images, video) is provided through multiple servers or service providers at different locations very near to the end-users that seek that content across different geographical locations (in the globe).

Q3. What are the most fundamental categories of cloud computing architecture?

A3. The most fundamental categories of cloud computing architecture are:

• System architecture
• Software architecture.

Q4. What is CDN?

A4. CDN stands for "*Content Delivery Network*" or "*Content Distribution Network*". This is a relatively new concept of distribution of heavy content that usually consumes substantial volumes of bandwidth and time to load to the websites requested by any user across the globe. CDN handles the heavy content in such a way that it improves the loading time of the websites and enhances the user experience on the websites.

Q5. Write down two important impacts of distributed cloud computing on cybersecurity.

A5. Two major impacts of distributed cloud computing on cybersecurity are:

- **Extended attack surface**—The attack surface is defined as the number of all possible points, or attack vectors, where an unauthorized user can access a system and extract data. The attack surface has expanded significantly due to the extensive use of distributed cloud computing services in public cloud domain. A huge network of cloud components with multiple interfaces, vulnerabilities, and bugs opens up a new attack surface for the hackers. The hackers have already become technically so advanced that they adopt numerous techniques to exploit the extended attack surface.
- **Complex network**—The distributed cloud has made the enterprise networks more complex, large, and difficult to manage. A big network, especially the distributed cloud network with IoT devices, is very difficult to manage due to numerous reasons such as lack of comprehensive regulation, rules, standards, protocols and security strategy.

Test Questions

1. How does Distributed Cloud Computing work?
2. Why do we need Edge Computing?
3. How does a distributed cloud architecture work?
4. Which industries are using the distributed cloud the most?
5. What is Big Data Processing?
6. What are the challenges of distributed cloud computing?
7. How does distributed cloud computing affect cybersecurity?
8. How does Software Defined Infrastructure (SDI) work?
9. What is a Content Delivery Network (CDN)?
10. What is the client/server architecture?

Chapter 8
Quantum Computing

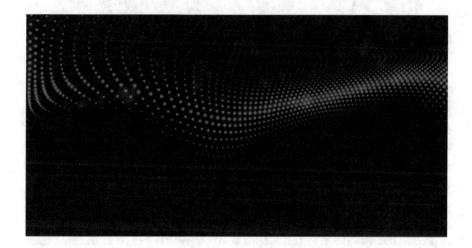

An Introduction to Quantum Computing

Quantum computing is a newly emerging technology in the field of computer science and technology. The quantum computing is based upon the quantum mechanics of an atom's particles. The theory behind quantum computing is quantum theory of mechanics in which multiple states of a particle of an atom is used. The most important states of a particle used in quantum computing include the following:

- Interference state
- Entanglement state
- Superimposition state.

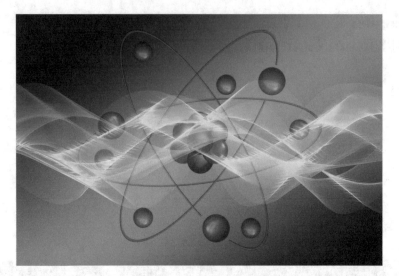

Fig. 8.1   Particles used in quantum physics/computing (*Pixabay*)

Quantum computing uses the quantum bit, precisely referred to as qubit, which is used as the basic unit of measuring the information or data in the quantum computing. In the classical computing systems, a bit is used for measuring the information or data. The bit is any one of the two states—either it is zero (0) or one (1). But in qubit, there is another state other than 0 and 1 called superimposition state, which is the combination of both states of a particle used in the quantum computing. A graphical representation of the states of a particular atomic particle is shown in Fig. 8.1.

In traditional computing, the transistors are used to process the data and increase computing power to the traditional computer machines while in the quantum computers, the physical particles of an atom such as electron and protons are used to process the data (used to solve a problem). A particle can be in multiple states in different conditions based on different physical processes applied to the atomic particles and the calculated superimposed condition, which is used as a state of that particular particle in processing the data or information to solve a problem.

Salient Features of Quantum Computing

Quantum computing is known for the capacity of processing that the people can achieve by exploiting different states of an atomic particle in different physical processes and environments. For instance, an atomic particle can exist in tens of states under the influence of different energy, charge, temperature, electromagnetic field, and other surrounding particles. The combination of those different states can make the quantum bit to have capacity of n-times exponential processing. The increased capacity can completely replace the existing (all) types of traditional computers such as supercomputers, mainframes, or other modern machines based on transistors or based on semiconductor materials. The quantum computers will use the superconductors and many latest materials used in quantum physics or quantum mechanics. The increased processing capacity can add numerous features, capabilities and characteristics to the quantum computers.

The main features, characteristics, and capabilities of this newly emerging technology are summarized as follows [209–211]:

- It is a subfield of quantum information science, which is one of the major domains of the modern quantum physics.
- Different types of computing models are used in building quantum computers such as:

  – Quantum Circuit model (mostly-used)
  – Quantum Cellular Automata model
  – Quantum Turing Machine model
  – One-Way Quantum Computer model
  – Adiabatic Quantum Computer Model.

- Maintaining the states of qubits is very sophisticated and difficult process because the states of the qubit after applying certain physical processes are highly unstable and prone to high level of state fidelity.
- Quantum computing is based on the principles of quantum mechanics in which the behavior and energy of particles are at atomic and subatomic levels.
- The power of qubits increases exponentially with the addition of another qubit. Normally, the power of qubit is 2^n with additional qubit numbers (n) unlike the traditional computing in which the additional transistor adds the power linearly.
- The hardware used in the quantum computers include superconductors and superfluid for maintaining and transporting the states of the particles within the quantum computer processors.

Short History of Quantum Computing

The quantum computing can be traced back from the discovery of the idea of quantum physics back in 1900 through 1925. But the domain-specific work dates back to 1980s when Richard Feynman and Yuri Manin proposed the development of quantum computers [212, 213]:

- **1985**—Construction of quantum logical gates by David Deutsch of Oxford University for building the processor of universal quantum computer. He is also known as the father of the modern quantum computing systems.
- **1994**—An algorithm to factorize numbers was built by Peter Shore.
- **1998**—First 2-qubit computer was built by three pioneer engineers Isaac Chuang, Neil Gershenfeld and Mark Kubinec.

- **2000**—A 7-qubit computer was developed by Emanuel Knill, Raymond Laflamme, and Rudy Martinez in the USA by using trans-crotonic-acid.
- **2017**—IBM announced the first quantum computer to be used commercially within the company's network and partners.

A large number of technology giants such as IBM, Google, Microsoft, Alibaba, D-Waves Systems, Intel Corporation, Nokia Corporation, HP, and many others are working on their respective quantum computing system building projects. At the time of writing this book, IBM is expected to launch commercial services based on quantum computer services by 2023 and Google by 2029 at large scale.

What Is Quantum Physics?

Quantum physics is a very important and modern part of physics. It deals with the smallest particles and units of matters and energy such as electrons, protons, neutrons, photons of light, waves, and their interrelationship. It provides the detailed insight into the micro-particles of matters and energy. The behavior of the smallest particles under the influence of energy, electromagnetic impact, and other forces and external sources that result in the change in the state of the smallest particles are also the parts of quantum physics.

Basically, the determination of states of the smallest physical particles in different conditions and under the influence of different energy sources helps create a quantum logical gate to store different states of a particle. Those states of a particle in relationship with other influences created by internal and external energy sources are used as the basic unit of storing and processing data or information unit in the quantum physics that makes a way for the development of modern quantum computer systems.

Theory of Quantum Computing

The theory of quantum computing describes the implementation of the phenomena of quantum mechanics into digital computation for solving highly complex problems very fast. The quantum theory allows the engineers to use multiple possible states of physical particles at atomic and sub-atomic levels to describe interrelated values by combining two basic states of an atomic particle [214].

For instance, our traditional digital computing machines understand the logical language based on the binary states in which the voltage or current is used in the state of either *present* or *absent*. If it is present, it means the logic (1) and absent means the logic (0). This limits the power of computing machines because there is no additional state that can be used for carrying out larger calculations in a faster way. The logical gates are used to store the states for carrying out calculations. To increase the capacity, additional number of transistors are required to added, which makes it much bulkier. The quantum mechanics uses more than two states of an atomic particle in movement, excitation, external source of energy to create multiple states of the particles for storing and processing the data. As noted before, the other states of 0 and 1 used in quantum mechanics include interference, superposition, and entanglement. The superposition states of 0 and 1 are used for achieving a large number of states for both 0 and 1 states (fundamental states of qubits). Thus, exponential growth in the capacity and power of the quantum computing machines could be achieved.

Working Principle of Quantum Computing

The working principle of quantum computing is based on quantum mechanics theory that helps achieve multiple states of a particle under the effect of energy and mass simultaneously. A particle can exist under the influence of multiple mechanical processes such as electromagnetic waves, temperature, flipping, rotations, and others. Multiple conditions or states can be achieved simultaneously under the impact of multiple quantum mechanics' influencem [215].

To grasp a better understanding of quantum computing, let us explain in analogy to the traditional computing. In traditional computing, logical gates are formed through transistor material that can hold two states at one time—either it is zero or it is one. The traditional computing can provide only four states due to binary systems; like, they can be 00, 01, 10, or 11. But in quantum computing, a particle can handle multiple states simultaneously under the influence of some triggers in the same space and time. Thus, you can achieve numerous states at the same time and space. You can achieve those different states by using the superposition effect that combines two states to get another one!

How Many States Are Used in Quantum Computing?

Theoretically speaking, the number of states that a quantum computer can handle is unlimited or infinite. The number of states is based on the qubit superposition conditions of a material used in the quantum computing. In most cases, the nucleus of an atom is used in which the basic qubit is same as 0 and 1 but a number of different states formed by the basic qubits can work simultaneously. This is not possible in traditional computing in which the processing takes place as one state at one time. But, in case of quantum computing, four possible states would work simultaneously and they would multiply with addition of another qubit to the machines. In this way the number of states will increase exponentially in terms of qubit states. In traditional systems, the number of possible states will also increase exponentially but they will process simultaneously. This is the major difference between the two types of computing systems. Thus, the number of states supported in the quantum computing is infinite.

As far as the present processors based on modern quantum computing are concerned, they can handle 2^7 states simultaneously. The exponential increase will continue with the addition of another qubit into the quantum computing system [216].

What Are Superimposition and Entanglement in Quantum Computing?

The superposition is the combination of two distinct states of quantum particles to form a new state that describes the linear combination of those two distinct quantum states. Thus, we can say that superimposition is a quantum phenomenon in which two states of a qubit are linked to gather to get a new state based on the linear combination of the two distinct states.

In quantum physics, the entanglement is an important phenomenon described as the state of linking of two quantum particles present at any distance away from each other. As per theory of physics, there is a link between two particles available at any distance in the space. Establishing a relationship between those two particles to develop a new state is called as the linkage or entanglement of those substances and the state developed is known as the superposition state [217].

Difference Between Traditional Computing and Quantum Computing

There are numerous similarities as well as differences between traditional computing and quantum computing. The basic concept of both computing systems is to process the information through machines made of materials that can handle different states and conditions. The main differences between both types of computing include [218]:

- **Processing material**—In traditional computing systems, the material used for data or information processing would be semiconductor materials that are utilized in manufacturing transistors and computer chips to handle logical data. In quantum computing, micro particles at atomic and subatomic levels are used to handle logical states of computations. Different types of physical material can be used for getting their atomic particles that can handle a large number of superposed states on the basic qubit states.
- **Data units**—In traditional computing, bits are used as the basic units of data measurements while in quantum computing systems, qubits are used.
- **States of units**—The states of bits and qubits are same—zero (0) and one (1). But, only one state of the bits can be processed at a time in traditional computing systems while in the quantum computing, multiple states of qubits can be processed simultaneously.

- **Scaling up**—The scaling of the capacity of the traditional computing systems is linear. It can add up to handle large capacities but processing power will add up in linear fashion like for instance, adding one computing unit to another computing unit will make just two computing units. In quantum computing, the addition of the qubits will multiply the capacity exponentially, which means if you add one qubit, the number of processing will double in relationship with the previous capacity.
- **Suitability**—The traditional computing systems are highly mature technology and can be used for solving a wide range of problems through the mature field of software development. On the other hand, the quantum computing systems are highly suitable for factorizing the numbers, which means that they are highly suitable for breaking security or other algorithms that are used in encryption and ciphering systems. The other applications that are very relevant for the quantum computing include data analysis, machine learning, artificial intelligence, and simulation applications. Again, the regular applications that we use in our daily-life may not be that suitable for the quantum computing systems.
- **Cost**—The traditional computers are not so costly as compared to the quantum computing systems, which are just new and have not yet matured in the public domain. It may take some time to really get those in the desired usable form.
- **Environments**—The traditional computers can be used in normal environments and with a moderate air-conditioned environment but quantum computers cannot be used in normal-temperature environments. They need highly cool environment created through superfluid and other unconventional ways of cooling to maintain the stability and certainty in the states of the atomic and subatomic particles. The classical computing systems can operate in almost all types of environments such as small radio interference, a bit high-temperature environments, different light environments, and others, while the quantum computing functioning is highly prone to the uncertainties and changes in any environment that may cause a little interference related to the above-mentioned factors.
- **Error rate**—The error rate in traditional computers is much lower than in the quantum computing systems due to its nature to get influenced by the external sources and interferences caused by different types of physical factors.
- **Operations**—The operation of traditional computing is based on the linear algebra while the quantum computing uses the Boolean algebra for operations (of the computing processes). The states used in the classical computing are discrete while those used in the quantum computing systems are continuous.

Real-World Quantum Applications

Quantum computing is evolving and has not got matured as yet. The potential of this technology is very huge due to the exponential growth in the processing power at very low power consumption and operational cost. Quantum computing is more useful for those applications that require huge processing power, especially in the field of research and development, mathematical calculations for satellite communication and other processing-power demanding applications. The most common possible applications of quantum computing may include [219, 220]:

- **Healthcare applications**—In healthcare research, especially development of drugs and testing those drugs through simulation of behavior of number of molecules in a particular module will become very complex and difficult for classical computers to handle. The increased processing power of quantum computers can be very useful in simulating and comparing processes in drug manufacturing. It can also be very helpful in diagnosis of numerous types of diseases by examining the behavior and multi-dimensional attributes of body at atomic and molecular levels to provide the deeper insight into the status of the diseases. Meanwhile, it can also be very useful in treatments, especially those that are based on radiology and similar kinds of techniques of treatment.
- **Cybersecurity**—High power quantum computing can be used in decoding the most complex ciphering and encryption methodologies easily. By applying quantum computing in security codes, especially for communication in information technology field, it is possible to break those codes easily. The positive applications would be simulating the techniques to avoid the deciphering of security codes. The ability of quantum computing to handle combinatorics calculations (that grow exponentially to learn the patterns and behaviors of different entities) makes it suitable for building numerous innovative processing-demanding applications in cybersecurity to simulate various security techniques and figure out existing security issues in the networks.
- **Artificial intelligence (AI)**—Artificial intelligence is a processing-power thirsty field due to big data analytics, understanding the unlimited world, figuring out activity patterns of masses, and many such computing-power consuming applications. The quantum computing can fill that gap of processing power in AI to pave the way for substantial growth of this field in different applications. It can make machine learning more efficient. Again, real-time data analytics and data-driven automated actions, object recognition, movement recognition, and may other processes can become more efficient and effective to produce accurate results.
- **Financial applications**—Banking, finance, Forex (foreign exchange market) trading, and trading of other commodities in real-time environment across the world are highly processing-oriented fields and they require huge processing power to make it possible in real-time and latency-free environment. The huge power of quantum computing can help numerous financial, banking, and trading applications work more efficiently and seamlessly.
- **Complex manufacturing**—The involvement of combinatorics or processing of multiple components in mathematical calculation is huge in numerous processes of manufacturing that are complex and require huge processing power. The use of quantum computing in such applications can make them more useful, reliable, accurate, and efficient. Any process that involves thousands of other processes where they are interrelated would require huge processing capacity and capability,

which could be easily provided by quantum computing. Thus, the application of quantum computing in complex manufacturing will prove so useful.

- **Telecommunication**—Major fields of telecommunication systems used in variety of domains such as defense, security, research and development, telemetry, meteorology, radar communication, satellite communication, and others will benefit hugely from the power of quantum computing systems because they need high-level and faster processing of signals to perform large calculations based on very complex algorithms. Quantum computing can provide that power to process highly complex and huge data/information in real-time environment.

Other than the above-mentioned domains, there are many other possible fields in which the use of quantum computer may prove very beneficial such as:

- Logistics and supply chain
- Digital marketing
- Software development
- Defense and national security
- Military research and development.

Major Projects on Quantum Computing

Quantum computing is proving to be a pathfinder technology in the modern world of emerging technologies due to its huge processing capacity and relatively low-cost operations. With the advent of modern communication systems and new range of information technology applications, the volume of data has increased significantly, which requires huge computer processing power. The traditional computing systems have been struggling in catering to that ever-increasing demand for the processing capacity. Quantum computing comes up as the savior. To cater to the increased demands for computing power and increased efficiency in modern processes, numerous big companies have already started their quantum computing projects. A few of them are mentioned below.

IBM

IBM is known as one of the pioneer companies in the field of quantum computing adaptation. The company announced its roadmap for quantum computing journey in 2020. Before the launch of that roadmap, IBM developed many components for the quantum computing systems. The experiment of launching quantum computing in cloud environment was tested in 2016 and 2017. The development of Circuit and Qiskit was one of the leading works done by the company. The starting point of IBM roadmap began with Falcon 27 qubits system module and it plans to reach 4000 + qubits processors by 2025. The company has developed numerous system modules during last few years. The development of Qiskit Runtime environment is one of the pioneer platforms for developers. In 2020, quantum algorithms and applications were developed along with quantum serverless platform. Many software applications to run in the cloud environment have also been developed. This project is targeted to cross 10 K qubit processing within next three years by 2026 (expected projection at the time of writing this book) [221].

Honeywell

Honeywell is another leader in the quantum computing market. It uses charged atoms (ions) for holding quantum data. In fact, it uses laser, microwave signals, and electromagnetic fields to hold the charges on the ions and manipulate the coding of data on those charges in the quantum systems, which are powered by highly reliable hardware and a range of operating systems and other software platforms. The traps use electromagnetic field for holding ions. It has two major projects related to quantum computing services known as:

- Honeywell Quantum Solutions (HQS)
- Cambridge Quantum Computing (CQC).

Nowadays, both have been merged to offer a comprehensive quantum computing-based solution for highly demanding industries such as defense, aerospace, cybersecurity, complex manufacturing, finance, pharmaceuticals, etc. The qubits used in Honeywell solutions are referred to as nature's qubit because they are defined in natural environment in such a way that they are the most stable and error-free [222].

Google

Google has a very well-crafted program for developing quantum computing system. The main objective of Google quantum project is to develop a 1,000,000 physical qubit computing system, which will help simulate behaviors of molecules to develop advanced medicines and many other inventions. The journey of Google's quantum project started long ago before it announced the supremacy of its project in 2019 when Noisy Intermediate Scale Quantum (NISQ) platform was announced. Then, simple chemical reaction was simulated to demonstrate one step forward in this journey. Later, Fermi-Hubbard model was simulated with 16 qubit computing system.

The company has established collaboration with many universities and research institutes to work for the error-corrected quantum system, which is the main objective of Google. The creation of an open-source library CIRQ paved the way for more advance works in this field. In the pursuit of achieving logical qubit, Google developed 100-qubit quantum system for further research towards error-corrected quantum computers. At present, a separate campus for research has been established that includes state-of-the-art hardware and data center infrastructure to work on this ambitious project to achieve the target of 1,000,000 physical-qubit by 2029. A range of software and hardware infrastructures have been developed for this journey such as [223, 234]:

- OpenFirmion Library
- TensorFlow Quantum
- CIRQ Python Library
- Quantum Virtual Machine
- QSIM circuit simulator
- QuantumLib Repository
- Many other quanta computing services.

All those tools have been integrated with modern state-of-the-art laboratory available for universities and research partners.

Microsoft

Microsoft invested hugely in the development of quantum computing systems. It has established Azure Quantum platform for building hardware and software tools required to develop topological quantum bit. For this purpose, it has developed many building blocks and tools based on Majorana Zero Modes. A quantum laboratory has been established in Denmark for developing hardware to be used for quantum project. The main features, capabilities, software, and hardware achievements of Microsoft Azure Quantum include [225, 226]:

- Quantum software development platform supports Q#, CIRQ, QISKIT, and other development platforms and languages.
- Integration and development of nanomaterials and quantum mechanics equipment
- Open-source platforms for contributing to the quantum development.
- Equipment testing labs.
- Cloud-based quantum development and business partner access for research and development works in this particular field.

Main Terminologies Used in Quantum Computing

Quantum computing is the most advanced computing technology available in the world at this point of writing this chapter. It uses numerous techniques, technologies, physical materials, chemical behaviors, and range of scientific principles to materialize the concept of modern quantum computing. As the name indicates, the core technology used in this modern computing is quantum physics or quantum mechanics, which is the study of physical behaviors of the micro-particles at atomic and subatomic levels. A stack of different processes, principles, materials, and techniques are used in this technology. A few of them are mentioned below [227].

Superconductors

Superconductors are the materials in element or compound forms that offer practically zero resistance to current and there is no magnetic flux field available in the material. There are certain physical materials that can be made superconductors by creating suitable environment of temperature and pressure to attain the superconductivity behavior of the physical materials. Numerous materials at very low temperatures become superconductor of electric current or electric charge in certain pressure environments. That point of temperature is known as critical temperature of superconductivity. About half of the periodic table elements can be made superconductors under certain temperature and pressure conditions. Superconductors can be divided into two major categories [228, 229]:

* Type-I superconductors
* Type-II superconductors.

Type-I superconductors are those that are normally used as conducting materials to build wires and cables in our usual applications related to electricity propagation. The examples of Type-I superconductors include mercury, lead, zinc, aluminum, sulfur, etc. The critical temperature of these materials ranges between 0.000325 and 7.8 °K.

Type-II superconductors are different compounds of lead and copper. They become superconductors at higher temperatures as compared to those type-I superconductors. A compound consisting of $HgBa_2Ca_2Cu_3O_8$ structure becomes superconductor at very high temperatures. The temperature at which this material becomes superconductor is 135 °K. The other major characteristic of this type of material is that the electromagnetic fields can penetrate into those materials but the same is not possible in Type-I superconductor materials.

Superfluid

Superfluid is a physical state of a fluid under certain conditions when it flows without losing any kinetic energy due to its characteristics of zero viscosity. There is no friction in this type of fluid and it can maintain its state of motion without requiring any external source of energy. At absolute temperatures, a few substances behave as superfluid materials and maintain their state of motion without any requirement of external energy. The most salient characteristics and features of a superfluid material are listed below [230]:

- Superfluid keeps creeping the walls of the container and maintains its motion without any external source of energy.
- Turning the container upside down will not affect the existing motion of the fluid that is contrary to the property of normal fluids.
- Some superfluid materials are compressible while many others not compressible at all.
- Normally, superfluid offers very high thermal conductivity; therefore, it is usually used in cooling systems of sophisticated equipment, especially for quantum computing systems.
- The main examples of superfluid material include:

  - Helium-3
  - Helium-4
  - Bose Einstein Condensates (Some not all)
  - Lithium-6 atoms
  - Atomic rubidium- 85
  - Atomic sodium.

- A few use cases of superfluid include coolant for high magnetic fields, detection of exotic particles, and quantum computing systems.

Quantum Mechanics

Quantum mechanics is a sub-class or a branch of physics that deals with the behavior of very small physical particles at atomic or sub-atomic levels. It is used to measure the physical movement, conditions, energy levels, effect of external energy sources, and many other related conditions at very minute particle levels such as atoms, electrons, protons, neutrons, photons, and others. This is on the contrary to the classical physics, which deals with the physical particles at macro levels to measure its positions, conditions, presence, projections, energy levels, and many other things. As far as the principles of classical theories of physics are concerned, they are much easier and intuitive to understand because we can see, feel, touch, and hear those behaviors through our senses in the world. But, quantum mechanics deals with the behavior of physical particles that are very difficult to understand because classical physical theories do not apply in certain conditions. In fact, they produce very strange and unique results under certain conditions, which may contradict the theories in the classical physics. That is why quantum mechanics is very difficult to understand.

The most common features and characteristics of quantum physics are listed below [231]:

- The quantum mechanics is mostly based on the mathematical calculations governed by the rules of uncertainty and probability.
- It is the base for numerous fields such as quantum chemistry, quantum computing, quantum science, quantum field theory, and others.
- It differs from the classical physics mainly in the following areas:

- Momentum, energy, angular momentum, and other system bound quantities.
- The objects in quantum physics are waves-particle duality while in classical physics, they are two different domains.
- Limits of how accurate detection of values can be achieved prior to measurement due to principle of uncertainty.

- Major names of scientists that contributed to this field of physics include Max Planck, Albert Einstein, Ludwig Boltzmann, Neil Bohr, Werner Heisenberg, Erwin Schrodinger, Max Born, and many other mathematicians who developed numerous mathematical formulas to calculate the behaviors of atomic particles.

Qubits

As we mentioned before, Qubit is the most fundamental unit of data in quantum computing systems. In classical computing systems, the basic unit of data is known as bit, which is a two-state representation of a piece of information. Either it is a zero or a one. In quantum computing, qubit is also represented by a zero (0), a one (1) and a third state that is linear combination of both states. The linear combination of both states of qubit is known as the superposition in quantum physics. Thus, the number of data bits stored in a qubit can be more than two (2). The superposition state of quantum particles is the point, which can be utilized to develop a computing system that can handle multiple superposition states under two states of qubits simultaneously.

The main characteristics, capabilities, and features of quantum bit (qubit) are mentioned in the following list [232]:

- Both states of qubit—zero and one—can exist simultaneously along with a huge number of superposition sets of states associated with both basic states, i.e., 0 and 1.
- The number of superposition states depends on the technology employed in the development of quantum bits.
- It is just a mechanical analogy to the digit bit used in classical computing systems.
- Quantum logical gates are developed through qubit states to form a quantum computer.
- The examples of qubits may include polarization of photon, spin of an electron, energy levels of electron, angular rotation of a particle and many others.
- The material used in qubit can be a range of particles in entangled condition such as trapped ions, atoms, molecules, photons, quasiparticles, and others.
- Interference through waves or movements of the particles used in the qubit can be used to add the number of states through interference—either by increasing the volume or decreasing the volume of energy or motion of a particle.

Quantum Logic Gate

Quantum logic gate, precisely referred to as quantum gate is physical analogous to the digital logic gate used in the classical computing systems. It is a model of quantum circuit building unit. In other words, it is the quantum circuit used in processing the qubit information in quantum computing systems. The main difference between quantum logic gate and digital logic gate is that the former is reversible while the latter is not reversible; thus, it can be reused in other circuits easily. The quantum gates are also known as unitary operators and are expressed as the unitary matrices in relationship with some basis. There are different types of gates used in modern quantum computing systems such as Pauli gates, Identity gates, controlled gates, phase shift gates, rotation operator gates, Hadamard gates, Swap gates, Toffoli gates, Fredkin gates, Ising coupling gate, Deutsch gates, Universal quantum gates, and few other combinations [233].

Quantum Counting

The number of possible quantum solutions for a problem is called quantum counting. It is referred to as the efficiency counting of the quantum system too. The quantum counting is the number of probable solutions to a search problem based on different Grover's search algorithm (discussed next), quantum phase estimation algorithm and others [234].

Grover's Algorithm

The Grover's algorithm is a mechanism of finding out the solution to the problem in unstructured data quadratically. It is the second most powerful algorithm used in the quantum computing systems. It was proposed by an Indian-American engineer, Lov Kumar Grover in 1996. The most important aspects of Grover's algorithm are listed below [235]:

- It is commonly referred to as Grover's search algorithm in quantum computing world.
- This algorithm utilizes superposition and interference states for improving the search speed much faster.
- This algorithm takes quadratically lesser number of steps to find out the unique property of an item from an unstructured data.
- If classical computing takes N steps to find out the value of an element in an unstructured data, the Grover's search algorithm can find out in $\sqrt{N}$ steps.
- It uses amplitude amplification technique to implement the search for solutions in computing the probable solutions.

- It is extensively used for searching the solutions to the most complex computing problems in quantum computing.

Shor's Algorithm

Shor's algorithm is a mechanism to find out the prime factors of an integer. The time taken to get the prime factors of an integer is about polynomial of the conventional computing time. For example, the time taken for factoring an N number (integer) into its factors is O ((log N)3) time and O (Log N) space is called as Shor's algorithm, which is extensively used in breaking codes. The main characteristics of Shor's algorithm include [236]:

- This is the most fundamental algorithm developed in the field of quantum computing.
- It was enunciated by an American mathematician Peter Shor in 1994.
- This algorithm is very effective in breaking the encryption keys used in the modern security systems such as RSA, Diffie-Helman key, and others.
- Shor's algorithm comprises of two parts referred to as reduction of the factoring problems to the order-finding and then it uses the quantum computing for order-finding through supportive quantum algorithm.
- The first part of this algorithm is also known as classical part and the second part is known as the quantum computing part.

Josephson Junction

Josephson junction is the combination of two superconducting material separated through a non-superconducting material to form a tunnel that can pass current without any need of an external source of energy. The Josephson junction is extensively used in quantum mechanics, especially in quantum computing systems. According to the research outcome, a superconducting electron can tunnel through the non-superconducting material placed between two superconducting materials. This junction is called as Josephson junction after the name of Brian David Josephson. This effect was predicted and Josephson demonstrated the current produced through non-superconducting tunnel placed between two superconducting materials. The current is known as supercurrent in modern quantum physics. Josephson was also awarded Nobel prize for this discovery [237].

Sample Questions and Answers

Q1. What are the most important states of a particle used in quantum computing?

A1. The most important states of a particle used in the quantum computing include the following:

- Interference state
- Entanglement state
- Superimposition state.

Q2. Describe Quantum Physics in brief.

A2. Quantum physics is a very important and modern part of physics. It deals with the smallest particles and units of matters and energy such as electrons, protons, neutrons, photons of light, waves, and their interrelationship. It provides the detailed insight into the micro-particles of matters and energy. The behavior of the smallest particles under the influence of energy, electromagnetic impact, and other forces and external sources that result in the change in the state of the smallest particles are also the parts of quantum physics.

Q3. What is Superimposition in Quantum Computing?

A3. The superposition is the combination of two distinct states of quantum particles to form a new state that describes the linear combination of those two distinct quantum states. Thus, we can say that superimposition is a quantum phenomenon in which two states of a qubit are linked to gather to get a new state based on the linear combination of the two distinct states.

Q4. Describe how quantum computing can play a role for Cybersecurity.

A4. High power quantum computing can be used in decoding the most complex ciphering and encryption methodologies easily. By applying quantum computing in security codes, especially for communication in information technology field, it is possible to break those codes easily. The positive applications would be simulating the techniques to avoid the deciphering of security codes. The ability of quantum computing to handle combinatorics calculations (that grow exponentially to learn the patterns and behaviors of different entities) makes it suitable for building numerous innovative processing-demanding applications in cybersecurity to simulate various security techniques and figure out existing security issues in the networks.

Q5. What is Qubit?

A5. Qubit is the most fundamental unit of data in quantum computing systems. In classical computing systems, the basic unit of data is known as bit, which is a two-state representation of a piece of information. Either it is a zero or a one. In quantum computing, qubit is also represented by a zero (0), a one (1) and a third state that is linear combination of both states. The linear combination of both states of qubit is known as the superposition in quantum physics. Thus, the number of data bits stored in a qubit can be more than two (2). The superposition state of quantum particles is the point, which can be utilized to develop a computing system that can handle multiple superposition states under two states of qubits simultaneously.

Test Questions

1. How does quantum computing work?
2. What are the salient features of quantum computing?
3. How did quantum computing develop?
4. Why is quantum physics important?
5. How do Quantum Computers deal with Superposition and Entanglement?
6. What is the difference between traditional computing and quantum computing?
7. What are the real-world applications of quantum technology?
8. How do superconductors work?
9. Why is Quantum Mechanics so important?
10. What is Grover's Algorithm?

Chapter 9
Tactile Virtual Reality

An Introduction to Tactile Virtual Reality

As the word tactile implies, it is about virtual information based on or pertaining to the touch sense. This is one of the most modern Virtual Reality (VR) technologies that deal with the touch senses in a digital environment. This is the technology that deals with touch senses by using different haptic elements such as motors, actuators, pneumatics, hydraulics, and other devices. It can create the senses of touch to a physical body in connection with the digital information or data that stimulates those elements to produce the desired output required for creating a touch sense of certain intensity and nature.

Tactile Virtual Reality (TVR) technology is the combination of two technologies—physical elements and digital information or data to form a touch sense in a highly immersive environment of virtuality. The most important features and characteristics of tactile virtual reality are summarized in the following list [238, 239]

- This type of virtual technology creates the sense of touch by combining the impact of haptic elements and digital information.
- It is the latest technology in the field of virtual reality ecosystem.
- Advanced artificial intelligence and machine learning technologies such as motion detection, behavior detection, pattern sensing, and others are used in building the virtual reality technology dealing with the touch senses.
- The foundation of tactile motion is based on haptic technology, which deals the stimulations that produce the sense of touch.
- The haptic devices that produce touch-sense include kinetic components and tactile sensors. The sensors sense the motion and other patterns of a human body and creates the instructions based on that information for the haptic devices to provide suitable feedback so that the most desirable touch senses are created.
- The touch senses are created through a tactile simulation with the help of haptic technology, which uses vibration, force feedback, temperature senses, and allied feedbacks to create a tactile virtual reality environment.
- The most important use of tactile virtual reality in modern industries is mainly focused on entertainment industry such as gaming and 3D visual entertainment.

Augmented Reality and Virtual Reality

Augmented reality, precisely referred to as AR, and virtual reality, precisely known as VR are two highly emerging technologies in the modern software and information technology-based industries. Both of these technologies are mostly used in combination due to their complementary nature to enhance the qualities of each other.

The superimposition of computer-generated images, audio, video, or other senses on the existing real-world environments such as images and other contents to improve the quality of the real-world ecosystem is known as augmented reality. This technology merges two different scenarios—computer-added and real-world environments to generate a suitable ecosystem that helps understanding the newly-generated environment more effectively and easily. Augmented reality can be used in improving images, videos, perception, sound and others [240].

The virtual reality (VR) is also a virtual technology in which the real environments can be visualized in the virtual objects such as products used in home renovation, clothing, and many other products. Those products can be added to the real environments to provide a detailed look of the products and the real environments. The products are used as 3D virtual items that are placed in your real environments such as building, home, or even on your body to try and choose the most suitable one.

As noticed that both AR and VR technologies are very common in many aspects; still, there are certain differences too. For instance, the augmented reality uses the real-environment and then additional effects are layered in the forms of layers to enhance the look of the environments. The added layers consist of the virtual objects to the real environment. On the other hand, in the virtual reality, the surrounding environment is completely virtual and computer generated imposed on the real objects to simulate the outlook of the virtual objects in the real environment.

History and Evolution of Tactile Virtual Reality

The history of tactile virtual reality starts from the idea created by the artists to generate illusion in the minds to feel more sensations in the arts such as paintings, pictures, and other media. Thus, we can say that the history of tactile virtual reality started with the imagination of the people to create illusions in the minds, i.e., can be traced back to centuries from now. But in the scientific way, the history of tactile reality starts with the virtual reality in the early stages. It can be traced back in the nineteenth century when the stereoscopic photos and views were created. The onward timeline is given below [241]:

- **1838**—Stereoscopic photos and viewers era is dominated by the creation of two dimensional images through stereoscope that would generate a sensational view. The first stereoscope was developed by Charles Wheatstone.
- **1839**—Development of View-Master for virtual tourism by William Gruber.
- **1849**—Creation of lenticular stereoscope by David Brewster.
- **1929**—Creation of "Link Trainer" by Edward Link. It was a commercial airplane training simulator model.
- **1930**—The idea of pair of goggles was enunciated by science fiction writer Stanley Weinbaum in his creation named as Pygmalion's Spectacles.
- **1950**—The creation of Sensorama by cinematographer Morton Heilig.
- **1960**—The development of Telesphere Mask by Morton Heilig. It was the first head-mounted VR display.
- **1961**—Development of first motion tracking display named as Head-Sight by two engineers named as Bryan and Comeau.
- **1965**—Development of the concept of ultimate display explained by Ivan Sutherland.
- **1966**—Development of *Furness Flight Simulator*, which is considered as the first modern invention of flight simulator. It was invented by Thomas Furness.
- **1968**—Creation of the first AR/VR head mounted display (HMD) named as Sword of Damocles by Ivan Sutherland and his student Bob Sproull.

- **1969**—The emergence of the concept of Artificial Reality through multiple projects in which the computer generated environments could interact with the people in it. It was developed by Myron Krueger.
- **1975**—Creation of VIDEOPLACE, the first interactive VR system by Myron Krueger.
- **1982**—Development of Sayre Gloves by Daniel Sandin.
- **1987**—The term Virtual Reality was coined by Jaron Lanier, founder of Virtual Programming Lab VPL company.
- **1991**—Creation of Medina's Mars Rover by Antonio Medina.
- **1993**—Development of SEGA VR headset.
- **1995**—Nintendo Virtual Boy, a VR enabled gaming console was developed.
- **1999**—The comprehensive simulation depicted in scientific fiction film named as "The Matrix".
- **2007**—Google introduced US Street View.
- **2010**—Google Street View becomes 3D.
- **2012**—The Oculus was started for funding and further development based on Oculus prototype developed by Palmer Luckey.
- **2016**—Many projects in mobile phones and other gadgets were started by numerous companies like HTC, Rift, and others.
- **2018**—Development of Half-Dome HMD (Head Mounted Display).

Since 2016 and onwards, numerous rapid changes are taking place in the field. The standalone systems are getting more grounds as compared to the mobile-powered virtual reality and augmented reality projects. The resurrection of tactile technology is gaining pace unprecedentedly.

Types of Virtual Reality

The virtual reality is an emerging technology, which is going through numerous changes and advancements. The new types and domains are very natural to be created in near future. At present, there are three main types of virtual reality technologies as listed below:

- Non-Immersive VR
- Fully-Immersive VR
- Semi-Immersive VR.

Let us have some insights into the three main types of virtual reality [242].

Non-immersive VR

Non-immersive virtual reality is the most basic form of virtual reality in which you act with virtual environments through some controllers and input devices but you are not a part of that particular virtual world. The virtual environment and your physical world are two separate domains interacting through just display and input devices to interact with the virtual world. The examples of this type of virtual reality include the gaming consoles, personal computers, laptops, and others. In those devices, you access the software applications such as games, videos, and other items virtually through display or screen and interact with them through different types of input devices such as mouse, gaming console, keypad, etc.

Fully-Immersive VR

This is the most powerful virtual environment in which you experience yourself in the virtual world and all interactions can be experienced in the virtual environment through multiple gadgets worn on your body. The user feels realistic presence and experience of virtual world and all activities happening in the virtual environment are focused on the user of the environment. To get the experience of such fully-immersed virtual world, helmets, hand gloves, body connectors with different types of sensors are required to interact with that setting and you interact with them like you are in the virtual world in reality. Every type of your reactions such as movement of body parts, movement of eyes, and other actions performed by your body are sensed and incorporated into the virtual world to manipulate the environment according to your behavior. Such environments are materialized through specialized environment powered by motors, hydraulics, and many other equipment and network of sensors. There are numerous examples of modern fully-immersed games, medical applications, and numerous other simulation systems.

Semi-immersive VR

As the name implies, it is more immersive than the first type of virtual reality and less immersive than fully-immersive virtual reality. Thus, we can say, it is a mixture of fully-immersive and non-immersive virtual reality. In this type of virtual reality environment, you interact with the environment through self-movement and other actions via computer touch screens or gamming console but not through any specific environment or tools made for the immersive experience of the user. The interaction with 3D videos, images, and games is the example of semi-immersive virtual reality. The most important capability and functionality that would fall in this category of virtual reality is known as Gyroscope. It is a virtual space fixed around the vertical axis of your mobile devices and the motion is controlled by moving the device on the horizontal axis. The latest applications also support multi-direction movement of the device to view the virtual world such as moving up, down, left, right, and at different angles.

Neurophysiological Tactile Measurement Techniques

The measurement of tactile virtual reality in our real-world environment is comparatively complex issue, which requires high-level of expertise, detailed algorithms, and powerful machines. The measurements are done through different techniques in our modern world. A few very important technologies that are commonly used to measure the tactile senses and convert them into machine formulas are listed below. These techniques help develop the output through a simulator for the desired entity or a person to feel.

- Electroencephalography (EEG)
- Magnetoencephalography (MEG)
- Functional Magnetic Resonance Imaging (fMRI).

These techniques are described separately in the following subsections.

Electroencephalography (EEG)

Electroencephalography is a technique used to measure the brain's spontaneous electrical activities for a certain period. This technique is extensively used in modern healthcare applications as well as modern virtual reality environment where the mental activities are also sensed and incorporated into the VR-powered environments.

Magnetoencephalography (MEG)

Magnetoencephalography, precisely referred to as MEG, is a technique to measure the magnetic fields created by the electrical currents in the brain cells. Other than the modern virtual reality environments, this technique is used in many fields such as healthcare, artificial intelligence, and others.

Functional Magnetic Resonance Imaging (fMRI)

This is another very important technique to measure the changes in the blood flow in the brain. This uses the magnetic resonance (MR) to measure the changes in the flow. The blood flow can be used to measure the brain activities. This technique can also be incorporated to build the most advanced virtual reality environments.

Somatosensation and Its Types

Somatosensation is a type of sense, which is commonly known as the sixth sense. It pertains to a range of different sub-senses. Our brain has a neural network that deals with the perception of touches. The most common sub-domains of touch senses that fall under Somatosensation category include [243]:

- Temperature sense—Thermoception
- Pain—Nociception
- Pressure, discriminatory touch and vibration—Mechanoreception
- Balance—Equilibrioception
- Movement and position—Proprioception.

All these sub-senses are extensively incorporated in the virtual reality environment creation in modern computer software systems.

There are two different types of somatosensation or touches—active and passive touches—as explained below.

Active Somatosensation

Active somatosenation or active touch is the type of sensing that involves proprioceptive and tactile sensing under the control of movement and balance. In other words, it is a voluntary, self-initiated, or intentional contact of movement [244].

Passive Somatosensation

Passive somatosensation also referred to as passive touch is a type of sense in which the stimulus of the touch is imposed on the body when the body is unaware of the stimulus as well as not in motion. It is not self-initiated.

Major VR Terms with Definitions

There are numerous terms used in the modern field of virtual reality and many more will appear with the advancements of this concept and technology. VR technology is evolving and emerging; so it is very natural that more new terms will appear in the coming days. A few most common terms used at present are explained below.

Head Mounted Display (HMD)

Head mounted display is small optical display that is integrated into eyeglasses and mounted on helmets to put on the scalp. The display is integrated with the imaging through modified eyeglasses that form the image on the eye's retina. This glass does not hinder or block the normal view of the user rather it superimposes the real-world view of the user over the virtual view of the virtual world. Some modern HDMs also incorporate motion senses [245].

Haptics

Haptics is a term that is used in virtual reality communication extensively. It is a form of communication through touch sense or applying force in the shape of thrust, vibration, and others. It is also known as kinaesthetic communication in 3D technologies.

360 Videos

360 videos are also known through different names such as spherical videos, surround videos, and immersive videos. These videos are recorded through omnidirectional cameras, which can record all directions of the field.

Interactive VR

Interactive virtual reality is a virtual system that allows the user to interact with the virtual environment through different types of tools such as tracking sensors, hand controller, and others. Interactive VR is also referred to as 360 VR. It is fully-immersive and interactive content that allows virtual environment to surround the user in it [246].

Stereoscopy

Stereoscopy is a process of enhancing the illusion of mind in an image by adding depth into the image. This is done through merger of two images to increase the depth of an image. The word stereoscopy comes from two words from Greek language that means hard/firm and view/see. In modern VR technology, the merger or superimposing of multiple images is referred to as stereoscopy.

4D Virtual Reality

The term 4D VR refers to the incorporation of physical world into the virtual world in modern computer-vision applications. The latest VR-powered gaming and other types of entertainment applications use the 4D VR technology to enhance the user experience in interactive gaming as well as other fields.

Field of View (FOV)

The field of view, precisely referred to as FOV in VR technology, is the observable environment or world through a single camera or an array of cameras at one single moment. The area that is covered under certain criteria for incorporating the visual data into VR applications is called as field of vision or FOV in computer-vision technology.

Image/Video Stitching

Image/video stitching is a technique that is used in enhancing the field of view (FOV) in VR-powered applications. In this process, multiple overlapping videos and images are connected or merged together to increase the field of view and improve the quality of the view simultaneously. This technique is extensively used in numerous kinds of applications such as street view, video surveillance, and other applications [247].

Simulator Sickness

The simulator sickness is the discrepancy in the motion between real-world environment and simulator environment for a simulator user that may result in vomiting, lethargy, uneasiness, headache, ocular motor disturbance, disorientation, and drowsiness [248].

Cave Automatic Virtual Environment

Cave Automatic Virtual Environment, precisely referred to as CAVE in virtual reality field, is a fully-immersive virtual reality environment with the help of projectors. This room supports multi-person, high-fidelity audio, high definition 3D videos in an immersive virtual environment. The rear projectors project the graphics through stereo systems to create highly immersive experience for the users. It is highly inter-active and superfast systems to respond with the help of highly powerful computers and software algorithms powered by artificial intelligence (AI), machine learning (ML), virtual reality, and other cutting-edge technologies.

Mixed Reality

As the name implies, the mixed reality or MR is the combination of two worlds, i.e. real world and virtual world to create a new world where both real and virtual worlds coexist. The creation of a new world combined with the real and virtual environments does not exist in any one of those worlds but they coexist in both virtual and real environments. That mixed world is referred to as hybrid virtual environment that consists of virtual reality (VR) and augmented reality (AR) simultaneously. To understand this point, let us take an example. In virtual reality (VR), the user is immersed into the virtual world through supported gadgets while the augmented reality takes places in physical world by overlaying one object upon the other one. However, in mixed reality (MR), both versions of realities take place simultaneously.

Real-Word Applications of Tactile Virtual Reality

Tactile virtual reality is getting strong roots in different industries across the global markets, especially in the fields of entertainment, training and education domains. According to the Fortune Business Insight projection, the global market size of virtual reality is expected to reach USD $227.34 billion by 2029 from just USD $16.67 billion in 2022 with a growth of over 45.2% CAGR [249]. This indicates that new domains and areas will open in near future as far as the applications powered by the tactile virtual reality are concerned.

There could be numerous real-world applications that are powered by the tactile virtual reality that are extensively used in different industries and domains. Some of the most important of those real-world applications are mentioned below.

Video Games

Video gaming industry has remained as one of the most attractive domains for the gamers since its inception in 1958, which is about seven decades now. The video gaming industry has come across numerous eras where many kinds of technologies influenced this industry. With the advent of modern software tools, the industry started growing exponentially. The latest factor that is influencing this industry the most is virtual reality (VR) and tactile reality (TR) that implement numerous factors like touch, smell, blow, breeze, and many others through different feedback systems.

Video games powered by the virtual reality are also referred to as virtual reality gaming industry, precisely VR-gaming. VR-gaming is a software application of a 3-dimensional (3D) virtual environment, which is created through VR software in computer-aided games. The virtual environments are superimposed on the real world environments to make user believe or create an illusion of being in real-world in which all elements of virtual world can be interacted through body motions and gestures including eye-blinking, arm-movement, leg moment, and the movements of other parts of the body. These could be enabled through specialized gadgets made for the virtual reality environments [250]. The formation of virtual environment for gaming is accomplished through different motorized systems that are controlled through different sensors and VR software applications, which interact and establish control within this ecosystem. The specialized rooms that contain screens, wearables, motorized feedback systems, make the players fully immersed into the virtual world with proper reaction from the elements of the environment that interact with the players.

According to the fortune business insights forecast, the global market size of virtual reality in gaming field is expected to reach USD $53.44 billion by 2028 from just USD $7.92 billion in 2021 with a growth of about 31.4% CAGR during the forecast period, 2021–2028 [251]. This huge growth will be driven by the immersive environment in which the players enjoy the gaming in three dimension (3D) videos. Players can interact with elements appearing in the video through different motions and actions, which are powered by the special VR gadgets like HMD, hand-gloves, gaming controls, and other wearables that are emerging in the markets.

The examples of numerous types of virtual reality gaming applications include Microsoft HoloLens, HTC hive, Oculus Rift, Google Cardboard, Microsoft Play Station VR, and numerous others available in the market. There are many wearables that are used in modern gaming applications powered by the virtual reality software.

Education and Training

Education and training is one of the top industries that benefit from the use of virtual reality in its environment. The simulated training or education for a particular topic, process, or procedure leave highly effective impact on the minds of the students or learners as compared to the traditional way of learning through description or depiction in the books. The use of videos has already impacted the education and training industry significantly. Now, the implementation of virtual reality environments into the training and educational processes just would revolutionize this industry.

According to the business research company projections, the global market size of virtual reality in education and training sector will reach about USD $32.94 billion by 2026 from USD $6.37 billion in 2021 with a gigantic growth of over 39.7% CAGR during the projected period [252].

The most common examples and applications of virtual reality in education and training sector include the following:

- Remote learning or distant learning applications
- Blended learning applications
- Process or skill development simulator applications
- Highly immersive environments to learn through experience
- Collaborative and interactive learning applications
- Digital mobile applications and learning games
- High-tech training applications
- Virtual field trip applications
- Experience-based learning systems.

The main reasons for choosing the virtual reality in education and training are the effectiveness, easiness, safety, and cost-efficiency of those applications. A few very important reasons are mentioned below [253]:

- It offers great user experience through immersive environment.
- Saves substantial cost and time due to numerous contributing factors.
- Offers better remembrance as compared to traditional education.
- Offers options to take a virtual tour of places, areas, environments, and much more.
- Provides greater student engagement through interactive, responsive and immersive environment of the applications.
- Simulators offer basic trainings for achieving better safety in applications like pilot training, car-driving, and other such skills that may become very risky without basic knowledge gained through training simulators for the most basic skills and experience.

Product Development

The use of virtual reality in manufacturing industry is increasing extensively for almost all types of product development lifecycles. All sizes and domains of companies in manufacturing sector can be benefited from the features and capabilities of virtual reality (VR) for their product development process. The product development process is one of the complex and costly processes in which the products are designed and prototypes are developed. Verification of product design in real-world environment leaves a big impact on the market value of that particular product. If the product does not leave good impression on the users when it hits the market, the product may be considered as a *failed product*.

Virtual reality allows us to experience and examine the design of the product in virtual environment like we observe it in the real-world (gives some perception at the least). This capability helps companies reduce the product design failure issues. The companies do not need to put extra money on designing a prototype for user experience and feedback. The VR technology allows one to design and prototype a product in virtual environment with full confidence to reduce the cost, time, and resources in the entire product development lifecycle. Many companies like Ford and other manufacturing companies are already using virtual reality for designing and prototyping their car models and other products.

The main advantages offered by the virtual reality (VR) technology in product development are [254]:

- Significant cost saving on the designing and prototyping of products.
- Using virtual product development suite helps reduce the time of development and increases the accuracy of the product.
- Design for manufacturing and assembly (DFMA) is a new ecosystem to accelerate and improve the manufacturing process.
- Virtual reality or VR-enabled systems can handle very complex designing and prototyping of products very easily.
- Easily accessible technology for all types and sizes of companies in manufacturing field.
- Provides real-life experience of the product in virtual environment well before the product comes into existence.

With this, we would end this chapter here. In the next chapter, we will look into the futuristic technologies.

Sample Questions and Answers

Q1. What is the main difference between AR and VR technologies?

A1. Augmented Reality (AR) and Virtual Reality (VR) technologies are very common in many aspects; still, there are certain differences too. For instance, the augmented reality uses the real-environment and then additional effects are layered in the forms of layers to enhance the look of the environments. The added layers consist of the virtual objects to the real environment. On the other hand, in the virtual reality, the surrounding environment is completely virtual and computer generated imposed on the real objects to simulate the outlook of the virtual objects in the real environment.

Q2. What is Fully-immersive VR?

A2. This is the most powerful virtual environment in which a user experiences himself in the virtual world and all interactions can be experienced in the virtual environment through multiple gadgets worn on his body. The user feels realistic presence and experience of virtual world and all activities happening in the virtual environment are focused on the user of the environment. To get the experience of such fully-immersed virtual world, helmets, hand gloves, body connectors with different types of sensors are required to interact with that setting. Every type of reactions such as movement of body parts, movement of eyes, and other actions performed by one's body are sensed and incorporated into the virtual world to manipulate the environment according to the behavior. Such environments are materialized through specialized environment powered by motors, hydraulics, and many other equipment and network of sensors. There are numerous examples of modern fully-immersed games, medical applications, and numerous other simulation systems.

Q3. What do you mean by Active Somatosenation?

A3. Active somatosenation or active touch is the type of sensing that involves proprioceptive and tactile sensing under the control of movement and balance. In other words, it is a voluntary, self-initiated, or intentional contact of movement.

Q4. Define 'Haptics'.

A4. Haptics is a term that is used in virtual reality communication extensively. It is a form of communication through touch sense or applying force in the shape of thrust, vibration, and others. It is also known as kinaesthetic communication in 3D technologies.

Q5. What is the purpose of an HMD?

A5. Head Mounted Display (HMD) is small optical display that is integrated into eyeglasses and mounted on helmets to put on the scalp. The display is integrated with the imaging through modified eyeglasses that form the image on the eye's retina. This glass does not hinder or block the normal view of the user rather it superimposes the real-world view of the user over the virtual view of the virtual world. Some modern HDMs also incorporate motion senses.

Test Questions

1. What is Tactile Virtual Reality?
2. What is the difference between Augmented Reality (AR) and Virtual Reality (VR)?
3. What are the different types of virtual reality?
4. How does Electroencephalography (EEG) work?
5. What is '*Somatosensorial*' and what are its types?

6. What is Haptics?
7. What is Interactive VR?
8. How does Cave Automatic Virtual Environment work?
9. What are the real-world applications of tactile virtual reality?
10. What are the main advantages of VR technology for product development?

Chapter 10
An Overview of Top Futuristic Technologies

What Is Futuristic Technology?

Futuristic technology is a very common term used in the field of information and computer technology (ICT). This term refers to the technologies that can change the way people live their daily lives in different domains such as businesses, social activities, governments, research and development, range of other industries and their processes. Those technologies may emerge in different fields such as manufacturing, aeronautics, information technology, electronics and communication, electromechanical engineering, construction and materials, and so on. The most important of

K. Thakur et al., *Emerging ICT Technologies and Cybersecurity*,
https://doi.org/10.1007/978-3-031-27765-8_10

all those industries and domains, information and computer technology is the most powerful and futuristic domain where a range of revolutionary things are expected to happen. Thus, it can be said that ICT and other fields of technology powered by the ICT technologies will be the major drivers of the futuristic technology sphere (as is the case even today).

If we look at the automobile and aerodynamic fields of engineering, the ICT is drastically changing the way they work even at the present time. For instance, the most complex and technology-oriented processes in the manufacturing of automobiles or vehicles is the designing and prototyping of the vehicles. The designing and prototyping of the car models, and their parts is being extensively done in the virtual environments powered by computer software and virtual reality (VR) environments today. Hence, we can expect what will happen after a few years or decades when almost all manufacturing processes will be automated and integrated into one operations room powered by the modern computer and information technologies (i.e., ICT's involvement will increase manifold).

Futuristic technologies are continuously evolving and the applications of those technologies in all major industries are being assessed and figured out. Many industry experts believe that upcoming or futuristic technologies that are going to revolutionize other industries and businesses can impact almost all kinds of existing industries and may bring forth new domains of industries in the future. The most important industries whose existing shape may change or a drastic transformation may occur are mentioned in the following list [255]:

- Computer or chip manufacturing industry
- Automobile and other manufacturing industries
- Transportation and aviation industries
- Medical and healthcare industries
- Bioengineering and medicines industries
- Aeronautics engineering and manufacturing
- Mobile and computer industry
- Telecommunication
- Software development industry
- Agriculture and environment engineering.

If we look at the major upcoming technologies that will impact the above-mentioned list of industries and domains and even other fields that are not mentioned here, we can see a range of upcoming futuristic technologies that are capable of impacting any one or more than one industries substantially. A few of those futuristic technologies are described in the following topic.

Top Futuristic Technologies

A large number of existing technologies that have either emerged in the market or are getting deeper roots in different applications across the industries have been described in separate chapters in this book. In this topic, let us focus on those technologies that are either getting ready to hit the markets or will hit markets in a few years or a couple of decades.

A few of those technologies that will shape the future in different industries are mentioned below:

- 3D Printing Technology
- 6G Technology
- Autonomous Robots
- Artificial Neurons
- Artificial General Intelligence (AGI)
- Mind Uploading
- Driverless Vehicles
- Infrastructure Hacking
- Regenerative Medicine
- Digital Twin (DT) Technology
- Programmable Living Robots
- Human Augmentation
- Intelligent Process Automation (IPA)
- Space Elevator
- Rotating Skyhook
- Light Sail.

If we look at the names of the above-mentioned technologies that are in offing, we can realize that they can revolutionize almost all processes, activities, and industries of human beings on this earth. A large number of present processes, products, and activities will vanish completely and numerous new processes, activities, and industries will emerge in the market place. The ways humans live their lives in the present age will also change drastically or will transform completely into new activities powered by the most advanced and futuristic technologies.

The details of the above-mentioned futuristic technologies are given with full explanation in the following sub-topics.

3D Printing Technology

Three dimensional printing, precisely referred to as 3D printing is a revolutionary technology that will change the shape of our world drastically. It will use different types of materials in a material manufacturer or printer like we add papers in the existing printers. The computer will make a design of the product in three dimensions

Fig. 10.1   3D-printed furniture (Flickr)

and the print command will be passed with the design data to the material printer. The printer will choose the material as per design and will print a 3D material based item, which will be added to the products manufactured in the factories. 3D-printed furniture samples are shown in Fig. 10.1.

For instance, you want to build a bolt, you need to create a design and give a command to the printer, which will take the melted iron from the connected furnaces and will print a bolt based on the design developed in the computer. Now, you need to just take that bolt and install on the product where you need that bolt. The automation may even interfere with the taking and installing the bolt to the new product. The bolt may automatically go to the assembly area and then it can be even automatically installed by a robotic hand or machine.

The most important features and characteristics of three dimensional printing are summarized here [256]:

- 3-D printing is also known as additive manufacturing.
- Objects are created from either digital 3D model or CAD (Computer-Aided Design) models.
- The entire process of manufacturing may include numerous stages such as material depositing, solidification, jointing, and others under the control of computer programs.
- The fusion of the material takes place in layers—layer by layer—by using the fused materials such as plastic, powder grains, liquids, and others.
- Earlier in 1980s, the 3D printing was considered useful for prototyping. That is why, it was used to be called as rapid prototyping too.
- Nowadays, it is being considered viable for the industrial production and referred to as additive manufacturing.

- The most advanced process based on 3D printing used in the start of the second decade of twenty-first century is Fused Deposition Model, precisely referred to as FDM, which uses continuous filament of thermoplastic material.
- The term *additive manufacturing* evolved in around 2010 and now, it is getting popular across the industries and manufacturing processes.
- Other alternative terms used instead of *additive manufacturing* include *desktop manufacturing* and *rapid manufacturing*.
- The main manufacturing processes in which the additive manufacturing is being used nowadays include stamping, hydro-forming, and injection modeling.
- The general principles of 3D printing include the following processes:

  - Modeling
  - Printing
  - Finishing.

- A 3D printer can be mono-material printer as well as multi-material printer.
- A range of additive manufacturing processes can be added into seven categories that are listed below:

  - Vat photopolymerization
  - Material extrusion or particle deposition
  - Material jetting
  - Powder bed fusion
  - Binder jetting
  - Sheet lamination
  - Directed energy deposition.

Short History of 3D Printing

The history of 3D printing dates back to 1940s when the idea of three dimensional printing came into the minds of the scientists. The chronicle history of 3D printing is described below:

- **1945**—The concept and process of 3D printing introduced by Murray Leinster.
- **1950**—Raymond Jones published this concept in science fiction magazine with concept of molecular spray under his article "Tools of the Trade".
- **1971**—Liquid Metal Recorder was patented by Johannes Gottwald.
- **1974**—The concept of 3D printing introduced by David Jones in New Scientist journal.
- **1980**—Two additive methods of fabricating 3D plastic models introduced by Hideo Kodama.
- **1984**—Computer Automated Manufacturing Process was patented by Bill Masters.
- **1984**—Alain Le Mehaute and his colleagues filed for patenting stereo-lithography processes.

- **1993**—Powder bed process using custom inkjet print head was invented by Emanuel Sachs at MIT Institute.
- **1995**—Selective Laser Melting process was invented by Fraunhofer Society.
- **2005**—An open-source project named as RepRap started by Adrian Bowyer. Under this project, a 3D printer was built in 2008 and the first replication named as "Darwin" of the newly built printer was achieved.
- **2009**—After the expiration of fused deposition meting (FDM) patent, numerous companies under RepRap open-source project emerged for FDM manufacturing.
- **2010**—Metal working process was accomplished first time. Before this, majority of the processes dealt with the plastics and polymer materials.
- **2016**—First LEAP engine with 3D printed fuel nozzles was manufactured by General Electric for Airbus corporation.
- **2021**—A fully 3D printed prosthetic eye was created and installed in a patient named as Steve Verze in Moorfields Eye Hospital London.

A large number of big companies in the field of manufacturing, healthcare, ICT, and other industries are investing enormously in 3D manufacturing. It is expected that many new processes and applications will emerge in the future marketplace very soon.

Applications of 3D Printing

There is wide range of applications of three dimensional printing in different domains of industries in the present day at a limited scale. But, it is projected that the applications of 3D printing would expand exponentially in all sectors of industries within a short period of time in the near future. The most important industries where 3D printing is being utilized at a certain level include:

- Manufacturing industry
- Fashion industry
- Firearm industry
- Healthcare industry
- Transportation and aviation industry
- Culture heritage industry
- Education and training sectors.

The major applications in those industries are mentioned in the following list:

- Work is going on to create 3D printed food items for astronauts and for the people on earth while maintaining the food texture and taste with the help of three dimensional bio-printers in the USA, Italy, and other countries. The plant-based meat developed in Italy is an example of such food items.
- Different companies like Nike and others are manufacturing clothes, shoes, and other items used in fashion industry with the help of 3D printing. Software-based systems create customized products for man/woman that are perfect for him/her.

- Numerous printed parts and components are being used in a range of products such as typhoon fighter jet, Airbus airplanes, Swedish supercar, GE aviation engines, US air force, and many other companies and their products.
- Different companies in the US are working on building 3D printed firearms. Those can be customized and used for special purposes.
- 3D printing based human body parts are being created at very small and experimental levels for implantation and other surgeries in healthcare industry.
- 3D-printed metal bridge was first time installed in Amsterdam.
- Numerous applications are being explored in the field of education and training, especially for the Science Technology Engineering and Mathematics (STEM) education systems.
- Numerous companies used 3D printing for manufacturing the PPE (Personal Protective Equipment) kits to cater to an exponential growth of demand during Covid-19 pandemic.

4D Printing

Four dimensional printing, precisely referred to as 4D printing, is an extended version or form of 3D printing, which is capable of printing a 3D object that can be morphed with the help of light, temperature, stress or other influencers with the passage of time. In other words, a 3D object printed through 3D printing technique can change its shape in a predictable manner controlled by the computer software program. The program exposes the 3D objects to different things such as temperature, water, air, electric current, magnetic energy, or any other source to achieve the desired impact on the three dimensionally printed objects.

A 4D printing can be done through single or multi-material to transform into different shapes from 1 and 2D that would strand into 3D shapes through a controlled environment governed by the computer program or software application interfaced with the printers [257]. First time, the word 4D was introduced in TED conference in MIT in 2012 and later on, the first research paper on 4D printing appeared in journal in 2013. The prospects of 4D printing are very high and it has potential to revolutionize numerous processes and industries in the future.

The notable companies that are using 4D printing for further research and development (R&D) as well as for different applications include [258]:

- Hewlett and Packard
- 3D Systems Corporation
- StrataSys Limited
- Autodesk Inc.

The demand of 4D printing will by huge in the future. According to a projection, the demand for 4D printing will remain over 42.95% after its launch in 1919–20 through 2025.

6G Technology

The sixth-generation technology, precisely referred to as 6G, is a mobile wireless technology that will be the successor of the 5G technology, which is still rolling out in many countries around the world. The 5th generation technology is considered as the most promising mobile data transmission technology that uses millimeter waves in spectrum of 24–50 GHz. The 6th generation cellular wireless network standard will use millimeter waves in range of 30–300 GHz and 300–3000 GHz too. The final development and testing of those bands and associated technologies are under research and experiments. Many companies are investing hugely to develop 6th generation technology faster to become the leader in the market. A few of those major companies that are investing substantially in research and development of 6G technology include [259]:

- Huawei Technologies, China
- Anritsu Corporation, Japan
- KeySight Technologies, USA
- Samsung Group, South Korea
- Ericsson Corporation, Sweden
- Jio Telecommunications, India.

The most important characteristics, features, and capabilities of the sixth-generation cellular technology network standards are listed below [259, 260]:

- The 6th generation wireless technology offers more throughput and much lesser latency as compared to the 5th generation networks.
- It is more heterogeneous or diverse network.
- The 6G technology supports more advanced ICT technologies such as virtual reality and augmented reality (VR/AR) in real-time environments, Internet of Things (IoT), mobile edge computing, diverse IT service-based business models, short packet communication, quantum computing and others.
- Uses upper spectrum in terahertz range to offer much more speed and data throughput or capacity.
- The 6th generation wireless cellular networks are targeted to offer speed at least 5 times faster than its predecessor 5th generation network. It may range above 100 Gbps.
- It is expected that 6G may support AI supported all programs in real-time environment such as driverless cars and other automated systems powered by artificial intelligence.
- The robust level of security, privacy, and secrecy may be the other most salient features of this technology.
- It may support up to one microsecond latency in different telecom services, which is even of much lower latency than 5G technology.
- This technology is also expected to fully support location awareness and present technology services for all its connected devices and networks.
- The releasing data of the 6th generation technology is set as 2030.

- Some experts and researchers believe that the peak data rate may reach one terabyte in certain ideal conditions or testing conditions.
- This technology is expected to further exploit wireless techniques used in the 5G technology such as beamforming, orthogonal frequency division multiple access (OFDMA), multiple input multiple output (MIMO), and higher sampling rates.
- It is expected to further expand upon the distributed radio access network (RAN) for achieving increased capacity and improving spectrum sharing.
- It is approximated that this technology will use wireless sensing solutions that will help the system to select the frequency band suitable for the given conditions.

Why is 6G Technology Needed?

Advancement in technology is the most fundamental component of modern businesses to create higher level of efficiency, reduction in cost, and improvement in daily business processes, and enhancement of service and product quality in the world. To achieve all those goals, a continual improvement in the technologies should be done (for all technologies). In modern world, the ICT has huge impact on all business processes and activities across the industries and sectors. Thus, the enhancement in the ICT technology is the most fundamental domain of research and development (R&D) for the improvement of all processes and activities in all industries. The demand for the number of connected devices and data bandwidth is increasing exponentially with every passing day. The present day technologies are not efficient enough to cater to that huge demand in the future. Therefore, a substantial investment in research and development is very necessary for the growth of the businesses and industries in the world.

The main reasons why our industries and businesses need the 6th generation technology are mentioned below [260]:

- **The convergence of technology**—There are numerous technologies that exist in the present-day wireless communication networks. The 5th generation technology is trying to coexist within heterogeneous environment and pave the way for larger convergence, which will be achieved through the power of 6th generation technology.
- **Incorporation of high performance computing (HPC)**—The existing technologies still lack in incorporating the high performance computing and quantum computing services, which are very fast and need high bandwidths and data channels for communicating with the similar kinds of devices on highly broader superhighways of data. The huge demand for low latency and faster speeds can be fulfilled through 6th generation wireless technology.
- **Internet of Things (IoT) network**—The number of IP-enabled devices are expanding exponentially, which would need huge bandwidth and high-capacity network connection for communication. The present day technologies will not be able to provide the required data after a few years or a decade. Therefore, a newer technology like the 6th generation technology is highly needed.

- **Mobile Edge computing**—The implementation of edge computing and mobile edge computing in a connected networks of billions of IoT-enabled devices would need huge data with very low latency. To make them capable of communicating in the real-time environment, the filtration and local provision of data are highly needed to avert any congestion in the network. This can be solved through edge computing systems and in the future, through mobile edge computing networks. Those networks require greater processing and bandwidth for real-time processing of the requests of billions of devices and other entities in the connected network. This can only be catered to with the help of newer technologies like the 6G technology and beyond.

Scope and Challenges of 6G Technology

The future scope of 6G technology is expected to be very bright in the light of modern technology trends and emerging technologies. A few highly data-bandwidth demanding services and technologies that will rule the future of the ICT businesses to provide support to the other industries include the following:

- Data centers
- Virtualization programmable networks
- Internet of Things (IoT)
- Quantum computing and high performance computing (HPC)
- Edge computing.

All of the above-mentioned domains of technologies and services are still not getting enough bandwidth to cater the emerging demands. Thus, the future of more powerful, low-latency, high-speed, and larger-capacity technology like the 6th generation cellular wireless network will be very bright and encouraging across the industries.

Every other technology or progress in any field brings a few challenges along with it. Hence, the 6th generation is also expected to face some very uphill challenges such as:

- Interconnection of a gigantic network of poles or towers that may bring complication related to public health, security, and cost factors.
- A challenge may be the security and threat detection in a huge network of diverse networks, especially separated through different edge devices, local wireless points and other nodes that provide local security.
- Implication in coordination between the core computing and edge computing that have to distribute certain processing levels and decisions at their own levels.
- Emergence of Nano-core that will deal with the processing of artificial intelligence (AI), high performance computing (HPC), and quantum computing.
- Such huge networks with huge volumes of network generated data will become another important challenge for the next generation networks of wireless technologies such as 6G. For the management of that big data, data analytics power

will be required. At the same time, the legal and regulatory issues will also emerge as a big challenge for the service operators.

- Governments and industries may face numerous other challenges in maintaining a range of activities such as law enforcement and social credit systems, health monitoring, air-quality and environment monitoring, monitoring sensory interfaces, which are the core components of the IoT-powered networks.
- And many others as the technology evolves

Autonomous Robots

Autonomous robots are intelligent machines that can perform tasks without any intervention of human being. They can perform the tasks based on the intelligence they possess through computer-vision training data sets. The autonomous robots are being used in different industries across the world. A few examples include the food serving robots, automated guided vehicles used in factories, flying robots or drones used for the surveillance purpose and may other domains such construction, environment monitoring, and so on [261, 262].

The main features, characteristics, and applications of an autonomous robot are mentioned below:

- It acts autonomously in the line of its learning through previous data fed into the software of the machine.
- No control or intervention of human in its core functions. A few critical maintenance or other functions may be in the control of engineers.
- The examples of autonomous robots include industrial robotic arms, self-driving cars, space probes, and flying robots.
- Normally designed to perform repetitive, dull, dirty, and dangerous tasks that are not favorable for a human to perform.
- The other areas where the autonomous robots are used in the present-day include e-commerce, warehousing, logistics, manufacturing, data centers, biotechnology, etc.

Main Components of Autonomous Robots

An autonomous robot works independently without any human help or human intervention; so, it requires the capabilities of thinking, perception, sensing, and decision making before taking an action through arms or other parts physically. Hence, the major components of an autonomous robot can be divided into logical and physical ones. The physical components may vary significantly in terms of its shape, size, and designs but the logical components of an autonomous robot will remain very much common. The most important logical components of an autonomous robot can be divided into three types as mentioned below:

- Perception

- Decision
- Actuation.

The perception part of autonomous component deals with the senses. For example, humans are capable of sensing different internal and external environment through five senses such as seeing, touching, smelling, listening, and tasting. Those senses help us make decisions based on the senses through our cognition that we learned through our experience in our life. On the other hand, autonomous robots would take two types of inputs known as perception such as:

- Proprioception
- Exteroception.

Proprioception deals with the internal sensing such as heating, force, batteries, charges, power, and other similar kinds of senses. While, exteroception deals with the sensing of the external environment such as sound, touch, light, smell, and many other objects. Both parts of the sensing are done through different sensors installed on the robots such as camera, bump sensors, torque-force sensors, spectrometers, thermometer, radar, and many other types of sensors.

The second important component of an autonomous robot is decision making. The decision making in autonomous robots is done through an artificial neural system powered by the computer algorithms. That decision making is done on the basis of the input that it gets in the surrounding and compares that with the authority limitations and assigned tasks or programs. The decision making process in human is done on the basis of gut feelings or brain's neural systems that would decide the situations encountered to make the most suitable decision (as deems to be). The human decision making can also be classified into two categories—gut feelings/brain decisions and the other one is the natural reflexes which just act fast on the basis of input without getting command from brain on the basis of its previous ready-made decisions such as blinking of eye, sensing of extremely hot environment, and similar types of things in high speed environments.

The third important component of autonomous robot is actuation, which is similar to muscle action performed by human body at the time when instructions are received from the brain after it makes the decision. The actuation in autonomous robots is done through different actuators that perform motion in different parts of the robots similar to the muscle movements performed by the muscles in human body. The actuators are formed through different equipment such as motors, hydraulics, electromagnets, and others.

Main Applications of Autonomous Mobile Robots (AMRs)

Autonomous mobile robots, precisely referred to as AMRs, are the most task-oriented self-operating and self-maintaining machines that can perform their respective tasks without any intervention or support from human beings. They are capable enough to tackle different issues such as obstacles and continue performing the designated tasks by overcoming those hindrances as much as possible. This is the reason that AMRs are being used in a large number of industries and different applications. The most common industries in which the most sophisticated AMRs can be used to achieve the most automated ecosystem include the following:

- Logistics industry
- Ecommerce and retail industry
- Manufacturing industry
- Warehousing industry
- Data centers management
- Biotech engineering
- Healthcare industry
- Research and development (R&D).

The most common applications in those industries are those that are very repetitive, dull, hazardous, risky, dirty, and odd jobs. The repetitive jobs can easily be performed by the autonomous robots such as taking one part from one tray to inset into a machine frame repetitively. A human can get bored and may not keep his/her focus on such repetitive job but a robot can perform it perfectly. Similarly, there are other jobs suitable for such robots, which are risky and dirty such as passing nearby running motors, furnaces, and other similar kinds of machines where human error can lead to risk to life of a human worker.

There are two major types of navigations performed by the autonomous robots in our modern industries:

- Indoor navigation
- Outdoor navigation.

The indoor navigation is much easier as compared to the outdoor navigation because indoor environment is limited and can be easily made understood to the AI-enabled machine through different training datasets in the modern machine learning applications. But, the external or outdoor environment changes significantly and machines cannot be fully trained for all types of objects, situations, and environments that encounter the navigation of autonomous robots. The indoor navigation can be made possible through a few sensors such as sonar sensing, CAD floor plans, camera, and other objects to make it understand the environment within limited premises. The outdoor navigation requires much sophisticated tools and learning algorithms and huge datasets to consume for building experience-based learning. An autonomous robot for outdoor navigation needs to deal with a range of issues such as:

- Huge disparities in surface densities
- Weather conditions and exigencies
- Three dimensional terrain to counter with
- Instability in the sensed environment
- And much more.

The prospective domains in the future where the autonomous robots will be consumed or deployed the most include:

- Space probes
- Military applications
- Automobiles
- Research and development (R&D).

The research and development work is going on in all major domains of industries, especially in space, defense, military, R&D, and other fields to make the most of the potential of autonomous robots.

Artificial Neurons

Artificial neurons are another very emerging and futuristic technology that will change the shape of automation and mechanization of all types of business processes with the help of robots, computer added programs, and other modern equipment and software platforms. Artificial neurons act similar to the biological neurons functioning in human brain. In human brain, the input data comes through multiple sources such as five senses and the gut feelings, which is considered as the sixth feeling, and then, all the inputs are processed to make a decision and pass the results to the next level if the neuron calculates that the weight of the output is such that it should be sent to the upper level or nodes or neurons.

The artificial neurons also resemble the biological neurons in structure, data processing, data input, and passing the output to the next level based on its weight. An artificial neural network consists of the following parts:

- Neurons or nodes
- Connections.

The neuron is a node that receives the data from multiple sources and processes it through certain logical algorithm(s) to generate the output, which is in the shape of weight or a voltage value. If that voltage value is above the threshold value (as set), the value is forwarded by the node to the next layer of neural network. If not, it is not forwarded to the next layer. There are two types of neurons in terms of their capabilities such as:

- Convergence neuron
- Divergence neuron.

The convergence is the capability of one neuron to collect input data from multiple neurons in the neural network. The divergence of a neuron is the ability of one single neuron to communicate with multiple neurons in the neural networks. The decision making in artificial neuron is done through activation function, which is built in the system and it runs on the basis of an algorithm. The algorithms of activation functions can either be linear or non-linear to decide the value in such a way that if it crosses the threshold value, it will be sent to the next layer and if not above the threshold value, it will not be forwarded. This process of deciding whether to send or not (the output value) to higher layer is known as "bias" in artificial intelligence or neural networks.

The main features and characteristics of a neuron are summarized below [262, 264]:

- Neurons are modeled similar to the structure of biological neurons.
- The assigned values are based on the importance of the input signal; it can be both positive and negative.
- The neuron's activation function works on the basis of hidden layer calculations to forward the result to output or not.
- The use of multi-layer perception or hidden layers in the future will help transform the ways people work, live, and act in the future.
- The advanced training methods to the artificial neurons or neutral network systems can help neurons understand and decide about the most complex environments and situations that may result in the most sophisticated automation in numerous industries, processes, and activities globally.
- The artificial or synthetic neurons are made up of silicon chips to process the input data through complex middle layer to generate an output.
- The process of learning about the environment in neurons is known as deep learning, which decides the outcome on the basis of importance of the input data.

As expected, the future of neural networks or artificial neuron technology is very bright in all domains of industries to achieve high level of automation, which is the future of highly connected world powered by multi-billion device networks.

Artificial General Intelligence (AGI)

Artificial intelligence (AI) is a very vast and futuristic technology, which is just evolving. The potential of artificial intelligence has not yet been fully achieved. In terms of levels of artificial intelligence capabilities, it can be divided into three generalized categories such as:

- Artificial narrow intelligence (ANI)
- Artificial general intelligence (AGI)
- Artificial super intelligence (ASI).

The modern world has not achieved the full capabilities of the first category of artificial intelligence known as artificial narrow intelligence. The most sophisticated artificial intelligent systems like IBM Watson falls in the category of narrow intelligence. The definition of artificial general intelligence is not standard but may experts define it as the level of artificial intelligence that matches the decision making capabilities of machines equal to human brains [265].

The artificial general intelligence (AGI) is also referred to as deep artificial intelligence, which is equal to the thinking, understanding, learning, and applying the intelligence to solve the complex problems in the way human brain thinks and updates the learned experience for the future applications so that the advancement or growth of mind/intelligence would continue. There are several companies such as Microsoft, OpenAI, Fujitsu, and many others, who are investing hugely in the research and development to achieve the AGI level of intelligence to change the world drastically.

Let us summarize the key features and characteristics of artificial general intelligence (AGI) that have not yet been materialized in the world [266].

- It is mid-level artificial intelligence that is defined as par to human intelligence.
- Still science fiction stuff; not materialized yet.
- It can handle abstract thinking, maintain background knowledge, use common sense capability, transfer the learning, and predict/find effects and causes of an event and much more.
- Machines would be capable of creating and correcting human work such as code, writings, and other similar kinds of tasks.
- Understanding human language and responding with the different types of languages by using the power of natural language processing (NLP) will also be a big characteristic of artificial general intelligence (AGI) machines.
- Navigation with the help of global positioning system (GPS) and motorized fine skills to carry out activities like human does will also become parts of the feature list of advanced AGI machines of the future.
- The other major activities that it can perform will include:

  - Utilizing different sets of knowledge
  - Handling numerous types of learning skills and learning algorithms
  - Creating the structures for almost all kinds of tasks that a human can perform
  - Dealing with metacognition and using it effectively
  - Understanding symbols and signs like humans do
  - Understanding of beliefs, ideologies, sentiments, and others.

As mentioned earlier, the AGI capabilities have not yet been realized. Many experts are skeptical of achieving such capabilities through machines and some scientists believe that it can be achieved by the end of the third decade of twenty-first century. One of the best scientists Stephen Hawking warned the humans not to create such level of artificial intelligence because it will be severely dangerous for the existence of human being. Even if that may not be exactly like what humans do for their understanding and action in a given context, the problem is that such machines can also err or there could be other issues like bugs in the code, mechanical failure or wrong electrical signal to machine parts, etc. that can create dangerous or hazardous situations.

Artificial Super Intelligence (ASI)

Artificial Super Intelligence, precisely referred to as ASI, is the most advanced form of artificial intelligence, which will supersede the capabilities and power of the most genius brains on the earth. This is just a science fiction or imaginary idea that will be developed in the future (as is the expectation or objective of the research community). The machines that possess the ASI-level power would be able to beat the most powerful brains available on the earth in numerous domains such as creativity, problem solving, decision making, approximation, and many other domains of human activities.

To have clearer understanding of the concept of artificial super intelligence (ASI), let us explore it in terms of different levels of artificial intelligence. All machines and software-based systems that can perform intelligent activities powered by the artificial intelligence (AI) fall in the first category or level of artificial intelligence, which is commonly referred to as weak artificial intelligence. The examples of weak intelligence based machines and software applications include driverless vehicles, expert systems, virtual assistants, and many other (today's) advanced AI applications.

The second level of artificial intelligence defined in the domain of modern information and computer technology (ICT) field is referred to as strong artificial intelligence, precisely "Strong AI". This is a software-based machine system whose capabilities and intelligence would be equal to that of a common human being. It can think, understand, make decisions, and accomplish approximations and self-adjustments, and many other functions performed by the present-day human brain. This level of intelligence is under research and development (R&D) and has not been fully materialized in the world as yet. May experts and scientists believe that it is near-to-impossible to create the intelligent systems of this level but efforts can be given in this direction.

The artificial super intelligence is the most advanced category of all other intelligent categories such as weak-AI and strong-AI. This is about the software-based intelligence systems that can surpass in all brain related activities that a human brain can perform such as thinking, decision-making, scientific creativity, social behavior adjustment, experienced-based wisdom, learning-based wisdom, approximation, adjustments in line with the surrounding conditions and sentiment readings of the people who interact with those machines.

The most important capabilities, features, and characteristics of ASI futuristic technology are summarized in the following list [267]:

- It is a fictitious concept of artificial intelligence that is based on imaginary suppositions.
- The software systems that surpass the capabilities of human brains is called artificial super intelligence (ASI).
- It is also referred to as "singularity" in the technological world and some scientists and experts refer to it as hypothetical artificial intelligence.

- If the concept is materialized, the ASI machines would think about abstraction and make interpretation of unclear things that the most genius human brains will not be able to think or crack (i.e., the hidden concepts).
- ASI-powered machines will be capable of not only reading of human emotions but also can evoke human emotions, ideologies, beliefs, and other abstract concepts.
- There are many philosophers and domain specialists that believe that the advent of the artificial super intelligence may lead to very risky and devastating situation for humans and other natural things on the earth or in the universe.

Mind Uploading

Mind uploading is another very futuristic and fictitious concept in the field of artificial intelligence (AI) and information technology (IT) that can transform the way people live not only on the earth but in some other parts of the universe too. The proposers of the idea of mind uploading believe that this process can be achieved by the mid of the twenty-first century when the artificial super intelligence or singularity will hit the markets.

According to the mind-uploading concept, the entire state of a human mind is scanned and transferred to the computer digitally so that the uploaded mind file can be installed or restored in any other body. This is called whole brain emulation—all emotions, creativities, skills, innovation, thoughts, experience and everything—into digital data or executable file and uploaded to the computer machine, which can be used for restoring, transferring, and installing in other bodies. Cyber channels can also be used to transfer the full digital mind!

The digital computer then runs the simulation of information processing of the brain in such a way that it will respond perfectly like the human mind does with all sentiments, emotions and other abstract concepts. This concept mind uploading (or, brain uploading) can be materialized by developing a range of tools and programs such as:

- Brain Computer Interfaces (BCI)
- Connectomics
- Information extractions.

The role of brain-computer interfaces is very pivotal in realizing the concept of super intelligence in artificial realm and brain/mind uploading systems. The mind uploading can be accomplished through two main methods.

- Copy the brain and upload to the digital computer. It is also called as constructive mind uploading or simple mind uploading. In this type of uploading, the brain is scanned for the information and features, then the information of the biological brain is mapped and finally, it is stored or copied to upload to the computers or any specific computational device used for this particular brain uploading task. In this process, it is possible that the biological brain may not be able to survive due to numerous processes of copying and pasting activities and sometimes, some major components may be destroyed for copying different parts of the brain. The entire copied and uploaded brain will survive in the virtual reality and simulated world.

- Copy the brain and delete brain through gradual replacement of neurons. This process is also referred to as gradual destructive mind uploading. In this process, the entire brain of human is deleted and replaced with the computer program that emulates the entire brain. After that process, the entire body of the human will remain under the control of the software program that was emulated and transferred to the human body through different interfaces.

The most important aspects, features, characteristics, and capabilities of brain/mind uploading include [268, 269]:

- The most important technologies that can pave the faster way for the mind uploading include superintelligence, brain-computer interfaces, virtual reality, information extraction from minds, and so on.
- During the process of scanning, mapping, copying, and uploading, a few parts of the brains or entire biological brain may get destroyed intentionally or may not survive due to the complex processes of copying, pasting or uploading.
- The copied brain can be installed or connected through BCIs to a humanoid robot, cybernetic, or biological body controlled through computer programs.
- This technology is also viewed as digital immortality because the brain does not die but it can change the body from one human to another or to a robotic body.
- It uses different technologies and techniques such as serial sectioning, computational complexity, brain imaging, brain simulation, brain scanning, brain mapping, and others.
- Some supporters of the brain uploading idea believe that it can be realized by the year, 2045 while the others believe that it may take much longer and may get materialized in the last decades of this century.
- The creation of cochlear (implant) is the first successful step-forward in building brain-machine interface (BMI) or BCI, which is designed to stimulate the brain cochlear nerve electronically to restore sense of sound that is hard to be heard.
- A large number of researchers and companies are investing in the development of modern BCIs for restoring the motor skills for restoration of brain damage due to stroke or accidents.
- For this process, powerful computers are needed; if the quantum computing materializes as per projections, then mind uploading can become reality much sooner than expected.

Driverless Vehicles

Driverless vehicles have already hit the market in many countries on trial as well as commercial basis. The full-fledged usage of driverless vehicles would change numerous human activities, business processes, and government rules significantly.

Once the driverless vehicle ecosystem takes roots, so many other emerging technologies would grow for the improvement of this system. The use of drivers would reduce significantly through more automation in the automobile and transportation industries. The shape of logics and supply chain would also change drastically.

The vehicles that can run between two destinations with the help of cameras, sensors, and artificial intelligence AI-powered software without any intervention of human is known as driverless vehicle. There numerous other terms used for this particular matter such as autonomous cars, self-driving vehicles, and robotic cars.

The main features, capabilities, and characteristics of vehicular automation through driverless car system are summarized below [270]:

- According to the standards developed by the Society of Automotive Engineers, precisely known as SAI, there are six levels of vehicular automation or autonomy such as:

  - **Level 0**—No automation
  - **Level 1**—Shared control or hands on with the automated systems
  - **Level 2**—It is called as the hands-off automation of car driving
  - **Level 3**—In this level of automobile automation, the eyes are taken off but human assistance may be required
  - **Level 4**—In this category, the eyes are taken off and mostly, it is not expected that a human would be required to intervene
  - **Level 5**—Steering wheel optional.

- The most common sensors used in autonomous cars include radar, GPS, Sonar, inertial measurement units, thermographic cameras, and others
- The driverless cars are expected to hit the market at larger scale in the late 2020s and early 2030s of this century.

With autonomous cars getting strong grounds, the artificial intelligence and machine learning will have a strong shot in the arm for more investment in machine learning applications in the future, which will lead to more advancement in futuristic technologies.

Infrastructure Hacking

The infrastructure hacking is not a futuristic technology but a futuristic risk or challenge to the modern world. It is being used as the weapon of the war in the new format of hybrid war. Highly sophisticated tools are being used by the hackers to take control of the core infrastructure of a country such as water supply systems, drainage systems, electricity grid systems, dam control systems, traffic control systems and many others. To counter those threats, highly futuristic technologies will keep emerging so that the emerging threats can be mitigated.

Regenerative Medicine

Regenerative medicine is another highly futuristic technology that will change our lives significantly and can open up new dimensions of human and animal lives. The regenerative medicine is a process of healthcare therapy powered by the futuristic technologies to replace or generate new cells, tissues, and organs in human body so that the damaged or teared organs or tissues can be generated within the body systems. For instance, in our body, there are numerous such processes that regenerate the tissues when they are damaged such as skin, bone, or muscles—the body regenerate those parts by developing new tissues. But, this process is not common in all cases. For example, if your heart is damaged, it does not regenerate or your brain is damaged, it is not redeveloped. Thus, the scientists and researcher are exploring the options through the potentials of technologies to create a therapy based in the regenerative medicines that can build new tissues, molecules, cells, or even organs.

The main features, capabilities, and characteristics of regenerative medicine are [271]:

- Replacing, rebooting, or rejuvenating the damaged tissues due to diseases, aging, accidents, or other things with the help of modern technologies (it is known as regenerative medicine therapy).
- Regenerative medicine does not involve the surgery or replacement of a human organ with another human organ as it happens in present-day medical procedures. This is a type of medical procedure in which the damaged parts would repair automatically at different biological levels.
- It deals with three level of regeneration—molecular, cellular, and tissue level generations.
- Many of the therapies concerning this domain of healthcare are under research and development, though a few such works as stem cell therapy and other small therapies are under limited use in modern medication.
- A few major types of regenerative medicine therapies include:
  - Immunomodulation therapy and artificial organs developed in laboratories
  - Tissue engineering and biomaterial therapies by transplanting lab-generated tissues
  - Progenitor cell therapies for blood, bone marrow, fat, muscle cells and others.

The impact of regenerative medicine powered by 3D printing and many other futuristic technologies will be very high on the life expectancy, immortality (If at all, in any form—as imagined. In reality, impossible for human beings as they are created.), and other factors that are not in control of the humans nowadays.

Digital Twin (DT) Technology

Digital Twin, precisely known as DT, is a concept of a new technology in which the physical object, assets, or even a process is presented into a virtual world through software program. This technology helps the industries save the prototyping cost and operational failures of products and processes significantly. There are three core components of digital twin technology that form a complete framework of digital twin technology such as:

- Physical product or object
- Digital product or object
- Connection between physical and digital objects.

The connection is mostly the information or data that flows between them to simulate a physical product or a process to produce the data that can be analyzed and used to make decisions about the physical products or processes. The digital data flows from physical world—object or process—to the virtual world—simulated digital environment. The processed and valuable information flows from the virtual environment to the physical environment for suitable decisions based on the information processes through the most cutting-edge technologies such as big data, artificial-intelligence-based software analytics, and other advanced tools as well as technologies.

There are certain types of this model-based designing technology that are mentioned in the following list:

- Digital twin prototype (DTP)
- Digital twin instance (DTI)
- Digital twin aggregate (DTA).

Moreover, in terms of integration levels, the digital twin technology can be further categorized into the following types:

- Digital model (DM)
- Digital twin (DT)
- Digital shadow (DS).

The other main characteristic and features of digital twin technology are summarized in the following list [272]:

- The core characteristics of this technology is the connectivity between multiple physical entities, which may be the processes, functions, physical objects, and others with the help of a range of sensors that send data to the virtual environment. The virtual environment provides access to the concerned entities to get the suitable information received after collection of data that is processed through the integrated software.
- It allows the homogenization of data and information under a unified ecosystem.
- This is highly programmable and smart environment in which the model can easily be modified and corrected with the help of software program modification.

- Digital twin technology provides you with the detailed digital traces of the events and activities so that you can easily trace the events and reasons of any kinds of malfunctioning in the physical environment with complete details.
- It is known for its modularity characteristics in which this technology fully supports the modular designs of the products and processes. Thus, it is much easier to scale up, scale down, modify, and perform other processes.
- It uses different stack of technologies such as 3D modeling, Internet of Things (IoT), virtual reality (VR), augment reality (AR), and many others.
- This technology can be used in future planning of the cities and towns, development of smart cities, industrial manufacturing, construction industry, automobile, healthcare, and many other industries.
- The concept of digital twin technology was introduced publically in 2002 by Michael Grieves.

The advancement of this technology will be very useful for virtualization of a range of processes and physical objects in the future to change the shape of the industries significantly.

Programmable Living Robots

Programmable living robot is a concept in which an artificial robot will be formed from organic material that can perform different activities as programmed in it. It can be like a human, or any other living thing that is under control of programmed software applications that would instruct the machine to perform certain functions and activities. This concept is still a fictitious one but a great breakthrough was achieved in 2020 when a self-replicating robot was made from stem cells taken from the embryo and heart of clawed African frog. The newly created robot is named as Xenobot due the origin of the robot from the clawed frog of Africa whose biological name is Xenopus Laevis [273, 274].

This development is being considered as the way to creating living robots based on the software program that may function like a living thing under the control of software program. The creation of programmable living robots is the conceptual advancement of regenerative medicine and mind uploading concepts. The combination of both concepts will bring revolution in scientific world and will transform the entire world into a new world with majority of the processes, activities, thoughts, and concepts drastically changed.

The other main characteristics, features, and capabilities of modern discovery of Xenobot self-replicating robot include:

- It was made from organic tissues under the supervision of AI program
- The newly created robot can live for weeks without any food
- It can move and swim in the given environment, which can be programmed as per requirements in the field of medicines or any other applications
- The programmable living robot can collect different stem cells to replicate its body and characteristics, which is an amazing thing of this robot

- These robots can be programmed to carry out numerous activities in human medicine such as programming cancer cells, cleansing of clogged arteries, repairing of birth defects, and regeneration of tissues after any kinds of trauma
- The replication of Xenobot was based on the kinematic self-replication, which is one of the concepts proposed by mathematician John Neumann in 1940.

The advancement of programmable living robot technology will open up new arena for a range of industries to use it in transforming numerous processes, activities, functions and even capabilities of existing objects and things.

Human Augmentation

Human augmentation is a new concept of technology stack that can enhance the natural limits of human capabilities, efficiencies, and characteristics by implementing a range of techniques powered by many modern technologies such as [275]:

- Artificial intelligence
- Robotics
- Bionics and prosthesis
- Brain computer interfaces
- Genetic editing
- Nootropics.

The human augmentation technology is also referred to as Human 2.0 in certain domains. This new concept aims to revolutionize the world as well as change the entire humankind and many other living things. This concept has a goal to break numerous boundaries based on religion, principles, ideologies, beliefs, and much more in the future. It is a fact that many such ideas are coined in the field but only few see the reality.

Intelligent Process Automation (IPA)

Process automation is an older concept prevailing in all industries and sectors of businesses. The automation started with the robotic or motorized automation in which a tasks-based activity was automated with the help of a motor or moving machine. With the passage of time, the programmable automation came into existence in which more complex activities were automated with the help of pre-programmed sets of instructions. There are four major types of automations used in the industries that are listed below [276]:

- Fixed automation or hard automation
- Programmable automation

- Flexible automation
- Integrated automation.

The intelligent process automation, precisely referred to as IPA, is an advanced version of automation that may fall in integrated automation but it is fully powered by the artificial intelligence and machine learning. This automation will work automatically by self-learning, understanding the condition, and making suitable decisions for actions. Thus, the intelligent process automation is highly advanced form of automation that has not yet been materialized at full scale in the commercial uses but research is going on to integrate the following technologies through software and hardware integration and deployment such as [277]:

- Machine learning (ML)
- Artificial intelligence (AI)
- Robotic Process Automation (RPA)
- Integrated software platforms.

In intelligent process automation, the robotic tools can be instructed flexibly through intelligent software applications that are trained for the environment to understand and make decisions and issue command to the robotic tools to act based on the decision that intelligent software makes. With the materialization of intelligent process automation in all domains of industries, the industries will get fully transformed into a new shape.

Space Elevator

Space elevator is a science fictitious concept which describes that an elevator can be built between planets and space or between the planets. This space elevator will provide the means for transportation between space and planet. The main objective of this elevator concept can be the launching of satellites and other space missions through it at a very low cost. There are certain issues that the researchers are trying to solve. The most important one of them is the material that is able to withstand such a huge weight and counter weight. The latest research by a Google team is focusing on the carbon nanotube (CNT), which is the strongest material under development. The structure of this proposed space elevator would consist of the following parts [278]:

- Base station
- Cable
- Climber
- Powering climber
- Counter weight.

Once this technology is realized, a huge transformation in extra-terrestrial communication and other processes will take place.

Rotating Skyhook

Rotating skyhook is a physical concept of a rope that can list the objects from the earth to the space for transportation purposes. The concept emerged in the mid of the twentieth century. The latest idea is that there would be a long tether on one-side and a small tether for counter weight on the other side. The longer tether would be attached to a rotating strong satellite and dropped to the earth-side. The small tether would be directed to the space for providing counter weight. Any object that is supposed to travel to the space such as spacecraft is attached to the long tether and flung out of the earth's gravitation force to provide a very economical carrier for travelling to space through this structure [289].

This structure is just in the research phase now. There are numerous challenges and issues pertaining to tether material, physical principles, structure operational issues, and much more. Once this concept of building a rotating skyhook is developed, the space industry will be revolutionized and numerous opportunities and arena will open up.

Light Sail

Light sail is a method of propulsion for spacecraft, which has recently been realized in the propulsion of space crafts. This method is designed to use the radiation pressure exerted by the sunlight on a huge mirror. This method is also known with other names in the field such as photon sail and solar sail. Electric sail and magnetic sail are other proposed forms of propulsion for the spacecraft to be launched in the future.

The first experiment of the use of this technology was done in 2010 when IKAROS spacecraft used this method of propulsion for testing purposes. According to the physical principle, when solar light falls on the mirror, it exerts pressure, which can be used for propelling thrust for navigational objects in the space [280]. The main successful missions of this revolutionary spacecraft propulsion system include the following:

- LightSail 1 Launched Atlas V rocket
- LightSail 2 Mission by The Planetary Society.

Sample Questions and Answers

Q1. What will be impact of ICT on automobile and aerodynamic fields of engineering in the near future?

A1. If we look at the automobile and aerodynamic fields of engineering, the ICT is drastically changing the way they work even at the present time. For instance, the most complex and technology-oriented processes in the manufacturing of automobiles or vehicles is the designing and prototyping of the vehicles. The designing and prototyping of the car models, and their parts is being extensively done in the virtual environments powered by computer software and virtual reality (VR) environments today. Hence, we can expect what will happen after a few years or decades when almost all manufacturing processes will be automated and integrated into one operations room powered by the modern computer and information technologies (i.e., ICT's involvement will increase manifold).

Q2. What is 4D Printing?

A2. Four dimensional printing, precisely referred to as 4D printing, is an extended version or form of 3D printing, which is capable of printing a 3D object that can be morphed with the help of light, temperature, stress or other influencers with the passage of time. In other words, a 3D object printed through 3D printing technique can change its shape in a predictable manner controlled by the computer software program. The program exposes the 3D objects to different things such as temperature, water, air, electric current, magnetic energy, or any other source to achieve the desired impact on the three dimensionally printed objects.

Q3. What is the key concept of mind uploading?

A3. Mind uploading is a very futuristic and fictitious concept in the field of artificial intelligence (AI) and information technology (IT) that can transform the way people live not only on the earth but in some other parts of the universe too. The proposers of the idea of mind uploading believe that this process can be achieved by the mid of the twenty-first century when the artificial super intelligence or singularity will hit the markets.

Q4. Define Digital Twin.

A4. Digital Twin, precisely known as DT, is a concept of a new technology in which the physical object, assets, or even a process is presented into a virtual world through software program. This technology helps the industries save the prototyping cost and operational failures of products and processes significantly. There are three core components of digital twin technology that form a complete framework of digital twin technology such as:

- Physical product or object
- Digital product or object
- Connection between physical and digital objects.

Q5. What is Space elevator? What is the main objective of this?

A5. Space elevator is a science fictitious concept which describes that an elevator can be built between planets and space or between the planets. This space elevator will provide the means for transportation between space and planet. The main objective of this elevator concept can be the launching of satellites and other space missions through it at a very low cost.

Test Questions

1. How can Futuristic Technology be defined?
2. Which are the top Futuristic Technologies?
3. How did 3D printing begin?
4. What are the applications of 3D printing?
5. In what ways does 6G technology differ from other technologies?
6. What is the need for 6G technology?
7. What is a data center?
8. What is an autonomous robot?
9. What are autonomous mobile robots (AMRs)?
10. In what sense are Artificial Neurons useful?
11. What are AGI and ASI?
12. Why is Digital Twin (DT) technology Important?
13. What is Human Augmentation?

Chapter 11
Impact of Advanced and Futuristic Technologies on Cybersecurity

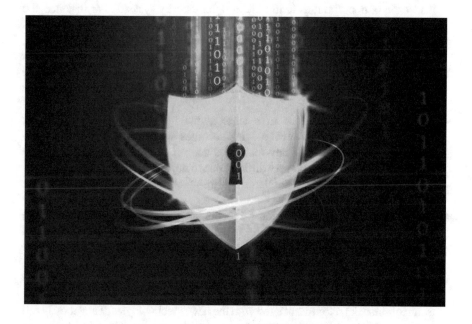

Overview of Impact of Modern Technologies on Cybersecurity

The threat surface expands with the expansion and utilization of more technologies and technological ecosystems because the new technologies that are evolving are used in the business much faster to achieve the competitive-edge in the market and to capture as much share of the market as possible. This race of capturing more markets with the help of advanced and emerging technologies often leave numerous security aspects skipped and overlooked. The hackers exploit the possible vulnerabilities and overlooked loopholes in the security of the software applications to hack into the security system and unleash cyber-attacks on a vast threat surface.

How does threat surface expand with the usage of advanced and emerging technologies? There are many aspects of using emerging technologies that can expand the threat surface. A few of those contributing factors include [281]:

- Emerging technologies have not got fully matured because technologies keep evolving with cybersecurity loopholes and then, exploitation from the hackers is noticed. Thus, at the starting point the cybersecurity threat landscape increases significantly due to those numerous security loopholes in different parts of the software and associated firmware.
- The number of intrusion points increase, which provides the hackers a large landscape of opportunities to intrude into the security systems. For instance, the number of devices connected under the technology stack of Internet of Things (IoT) is in billions. All those devices use different types of firmware, software, communication protocols, and security mechanism. That provides a much larger area for hackers to exploit any vulnerability in the codes and mechanisms of those devices to get information about the other connected networks and devices for launching cyberattacks.
- Operations and maintenance of a large number of devices connected under the umbrella of emerging technologies becomes so difficult. The upgradation, malfunctioning, removal of any malicious code installed on those devices and technologies points become very difficult due to numerous factors. This increases the opportunities for cyber threat actors to attack the security systems of the technological landscape.
- The use of weak communication protocols such as signaling system 7 (SS7) and a few older systems that are still in use in many devices for communication increase the threat levels in the modern cybersecurity systems powered by the latest technologies.
- The operators of new technologies or the end-users (who are normally not expert users) incur troubles. The general users are often not the trained users who are aware of the risk factors associated with the newer technologies. They mostly fall prey to the hackers through numerous ways such as social engineering, password management, and service/product using behaviors.
- Compatibility and interoperability issues also increase with the deployment of modern and emerging technologies with the existing technological ecosystem. Different standards and operational guidelines may not be compatible with the other systems that create a disharmony and loopholes in the comprehensiveness of the cybersecurity of systems.
- Advanced skills and expertise of hackers who always try to explore the possible vulnerabilities in the emerging technologies often hit the markets. They are highly specialized and sophisticated people to figure out the possible chances of breaking

the security of the newly introduced technologies. This also increases the threat landscape substantially.

- The usage of advanced hacking tools such as botnets, network mapper (Nmap), Nessus, NetStumbler and many others powered by the modern technologies such as extended reality, machine learning, artificial intelligence and, any other mechanism make the new products and services more prone to the data breaches and other anomalies.
- The extended use of social media is also increasing the risk of cyber-threats because modern software that can analyze the user behaviors and their interests and their working patterns can convert them into different traps and use those patterns for hacking purposes.

Thus, the threat space expands with the usage of newer technologies and bigger networks. Finding the security experts or operations and management specialists for the newer or emerging technologies is often an uphill task. This also leads to expansion of threat landscape in the cybersecurity. Thus, the expanding threats landscape in the technological ecosystems powered by the modern/advanced technologies can be reduced or averted in certain cases by using the following techniques and schemes in the cybersecurity systems [282, 283]:

- Continual monitoring and updating of the software and hardware tools used in the networks through automated as well as manual schemes.
- Training the company staff as well as the users for maintaining as much security as possible to reduce the risk of cyberattacks.
- Use advanced technologies in communication such as transmission and cloud encryption for data storage and transportation.
- Usage of highly secure technologies in cybersecurity systems such as blockchain, artificial intelligence, machine learning, and others.
- Deploy behavior analytics powered by big data and AI analytics.
- Incorporation of context-aware security for making data driven decisions regarding the emerging threats in real-time environment.
- Implementation of defensive artificial intelligence (AI) powered tools for preemptive measures before the cyberattack can hit the assets.
- Deployment of Manufacturer Usage Description (MUD) standard for the security of IoT devices. This standard is built by the Internet Engineering Task Force (IETF) for home and SMB (Small and Midsize Business) automation.
- Incorporate the Extended Detection and Response (XDR) systems for preventing any emerging intrusion from the hackers.
- Adopt zero-trust or never-trust and always-verify policy in your company, partners, and stakeholders simultaneously. Deploying an architectural framework named as Zero Trust eXtended (ZTX) is the best option to implement this policy.
- Avoid using the faulty protocols in your communication systems such as SS7, Diameter, Session Initiation Protocol (SIP) and others that have already been exploited by the malicious actors in the cybersecurity field.
- Implement server-less cloud and container ecosystem for more robust cybersecurity in your system to prevent malicious intrusion or data breaches.

With the advancements in the technologies, the threat landscape increases in cybersecurity; at the same time, it offers numerous solutions to overcome the emerging cybersecurity challenges in the modern IT world.

Major Cybersecurity Challenges Due to Advanced Technologies

According to the Strategic Market Research forecast, the global cybersecurity market is expected to reach a whopping USD \$478.68 billion by 2030 due to huge impact of the emerging and disruptive technologies that have already shown their impact on the industries across all domains. The global market of cybersecurity in 2021 was measured as much as USD \$216.10 billion. This huge or exponential growth in the global market size of cybersecurity was estimated at about 9.5% CAGR during the forecast period [284]. The main technologies that have increased the landscape of cybersecurity threats include bring your own device (BYOD) standards and IoT technology ecosystem. Other emerging technologies that have also empowered the hackers and malicious users include artificial intelligence (AI), machine learning (ML), data analytics, big data, natural language processing (NLP), and many others.

The impact of those technologies is felt in numerous day-to-day personal, business, and governmental activities. For instance, the total cost of cybercrimes in the world is expected to cross USD \$10.5 trillion annually, which were just USD \$6 trillion in 2021 [285]. Looking at this staggering figure, everyone can understand the gravity of the cybersecurity threat to the businesses across all domains and industries in the world. The average growth in the global losses to the business world due to cybercrimes is expected to remain above 15% year over year for the projected time period.

There are domains other than business world that face serious challenges or threats due to the cybersecurity issues emerging through the newly introduced technologies and those futuristic technologies that are expected to hit the markets very soon. The strongest impact of cybersecurity due to the emerging technologies is especially noticed in the following domains:

- Risk to National Security
- Breach of Privacy
- Increased Burden of Cybersecurity on Businesses
- Shortage of Cybersecurity Specialists
- Risk of Extensive Data Exposure
- Society and Business Manipulation.

Let us now separately explore the above-mentioned domains where the impact of cybersecurity due to the emerging technologies is felt significantly.

Risk to National Security

Some quality research works on the impact of cyber threats on the national security suggest that cybersecurity threat not only impacts the businesses and people living in a country but also the national security of that particular country in different forms and manifestation. Today's world is so connected and interlinked that nobody can escape the impact of one thing in a country or even in a foreign country in terms of numerous factors. The impact can be divided into two separate categories—(1) that, which is inflicted on a nation as direct cyberattack on the security institutes, agencies, and their assets—(2) that in which the entire ecosystem of a nation is disturbed, which results into chaos and internal disturbances.

The main impacts of a cybersecurity that can lead to national security issues include the following [286]:

- Loss to major businesses that contribute significantly to the economic development and stability of a nation.
- Stealing and manipulation of data of the common people and target them for the activities that can be detrimental and destructive for the country's security.
- Creating panic and social disturbances through different means when the data of the most vulnerable sections of the society is available for exploitation.
- Stirring political chaotic conditions based on the sensitive data and information collected through data breaches and cybersecurity attacks.
- Tearing the social fabric and community harmony in any country for destabilizing the economic, political, and social activities.

The other impact of cybersecurity on the national security may be very direct, in which the cyberattacks are launched on the data, secrets, systems, assets, plans to steal and misappropriate them in accordance with the needs of the adversary that wants to inflict damages to the national security a particular country in certain conditions. Another direct impact can be the cyberattacks on the utilities and other civic services so that the chaotic conditions are created in a country to carry out the most dangerous designs and activities by exploiting those conditions.

The innovative and emerging technologies add fuel to this impact because those technologies are released early (than the time of maturity) and often to establish competitive-edge by compromising the impact of the cybersecurity on the businesses, national security, and the people of the country. Thus, a serious warning should be taken by the concerned businesses, governments, institutes, and people about the impact of the cybersecurity on the national security, which is the most fundamental component for running governments, businesses, and societies smoothly.

Indeed, the threats to the states and communities exist through the cyberspace. Therefore, cybersecurity should always be treated from the national security perspective and standpoints.

Breach of Privacy

According to the latest information, the total number of people that were affected in breach of privacy impact in the US only was 53.35 million in just first quarter of the 2022. With this figure, it is not very difficult to imagine about the impact of data breach on the people across the globe [287]. Almost, every Internet-connected or online person has shouldered the impact of privacy breach in one way or the other. The breach of privacy keeps increasing with the emergence of the modern technologies due to the same reasons such as:

- Over excitement of the users to use the newer technologies without knowing or taking care of the impact of those technologies on privacy breaches.
- Early release of products or services based on the innovative technologies by the companies while ignoring the privacy breach issues.
- Unavailability of domain expertise and security professionals to deal with the emerging data breach threats.

Increased Burden of Cybersecurity on Businesses

The scale of damages caused by the cybersecurity threats in the world is huge in trillions a year. To avert or reduce the impact of that huge devastation, businesses focus on investing hugely on the cybersecurity to maintain a robust and reliable security level to avoid any kinds of damages caused by the cyber attacks. The average annualized spending by the businesses in the cybersecurity field is expected to cross USD $1.75 trillion by 2025 [288].

This huge burden is increasing very fast due to the emergence of the most advanced and latest technologies and innovative platforms. Thus, the businesses are badly impacted with the burden of such as huge amount of money, which is spent on the maintenance and enhancement of the cybersecurity systems and professionals.

Shortage of Cybersecurity Specialists

Another impact of emerging technologies is the shortage of the cybersecurity professionals. Entire world has already been walking through tough time due to the shortage of skilled and qualified cybersecurity professionals. The advancements in the emerging technologies and newly introduced technological business ecosystems have aggregated this problem for all types of businesses and government agencies simultaneously.

According to the latest estimation by World Economic Forum, there is a huge shortage of cybersecurity professionals in the world. There are about 3,000,000 experts and specialists short of the requirements in this field. This demand is continuously increasing and the availability of the cybersecurity professionals is continuously depleting. Thus, the impact of modern technologies on different entities—business and governmental—is the shortage of cybersecurity professionals, experts, and specialists [289].

Risk of Extensive Data Exposure

The most competitive business environment is where gaining competitive-edge through different innovative ways is the only way for the companies and organizations. In such environment where the hackers and malicious actors are more organized and bold to attack and there is a huge shortage of cybersecurity professionals to counter them, the chances of extensive data exposure will remain very high. With the excitement of newer technologies among both the users and providers, the situation is becoming more precarious for extensive data exposures and data breaches.

The main reasons of data exposures pertaining to modern and innovative technologies include the following:

- Use of huge number of devices in highly diverse environments of IoT where a large number of devices run diverse firmware and software.
- Increased number of user accounts with those huge number of devices and related services are also prone to data exposure.
- Mismanagement in password creation, maintenance, and storage by the clients.
- Outdated software and devices that are not updated regular also exaggerate the situation.
- Continual emergence of innovative techniques and ways that people are not fully expert at.

Society and Business Manipulation

Another impact of the modern and emerging technologies on the cybersecurity is that by breaching the control of cybersecurity or through other means, the entire society of a country or a business ecosystem of a nation can be manipulated and streamlined for the benefit of the adversary. This can be done through the most modern and innovative technological approaches to analyze the behaviors and activities of the people and businesses in a particular field or industry and achieve the most actionable data that the people and businesses are influenced with.

The examples of such incidence were reported in the US presidential elections in 2016. In those elections, it is alleged that Russia tried to manipulate the mandate in favor of a particular party by building an opinion in the people in such a way that they would be easily influenced. The similar kinds of allegations surfaced in India too before 2019 elections in the country. The social media, print media, online news, and different news channels can be used as the tools to implement the agenda based on the data achieved from the target audience or a society.

Hence, cybersecurity aspects span a lot of different subjects in today's world. With the emerging and innovative technologies, we have been benefited as well as new threats and risks have emerged. Meticulous studies could eventually solve issues in the coming days and there is indeed no alternative to continuous research in these domains.

Sample Questions and Answers

Q1. Name some domains where the strongest impact of cybersecurity is noticed due to the emerging technologies.

A1. The strongest impact of cybersecurity due to the emerging technologies is especially noticed in the following domains:

- Risk to National Security
- Breach of Privacy
- Increased Burden of Cybersecurity on Businesses
- Shortage of Cybersecurity Specialists
- Risk of Extensive Data Exposure
- Society & Business Manipulation.

Q2. What are the reasons for increasing trend of breach of privacy with the emergence of modern technologies?

A2. The breach of privacy keeps increasing with the emergence of the modern technologies due to the following reasons:

- Over excitement of the users to use the newer technologies without knowing or taking care of the impact of those technologies on privacy breaches.
- Early release of products or services based on the innovative technologies by the companies while ignoring the privacy breach issues.
- Unavailability of domain expertise and security professionals to deal with the emerging data breach threats.

Q3. Why is it difficult to find specialists in cybersecurity when modern emerging technologies are considered?

A3. Most of the users are general users or non-experts. When new and emerging technologies come to the market, most of the people are not aware of the risks involved with them. Even the existing experts need training regarding use of the technology and many cannot simply switch to the new mode of operations. Another factor is that their demand becomes high. Hence, it is often difficult to find specialists in cybersecurity in such situations.

Q4. Write down the main reasons of data exposures pertaining to modern and innovating technologies.

A4. The main reasons of data exposures pertaining to modern and innovative technologies include the following:

- Use of huge number of devices in highly diverse environments of IoT where a large number of devices run diverse firmware and software.
- Increased number of user accounts with those huge number of devices and related services are also prone to data exposure.
- Mismanagement in password creation, maintenance, and storage by the clients.
- Outdated software and devices that are not updated regular also exaggerate the situation.
- Continual emergence of innovative techniques and ways that people are not fully expert at.

Q5. Give examples how the society can be influenced by information manipulation over the online platforms.

A5. A good recent example could be the US presidential elections in 2016. In those elections, it is alleged that Russia tried to manipulate the mandate in favor of a particular party by building an opinion in the people in such a way that they would be easily influenced. The similar kinds of allegations surfaced in India too before 2019 elections in the country. The social media, print media, online news, and different news channels can be used as the tools to implement the agenda based on the data achieved from the target audience or a society. For achieving the goal, all types of online platforms can be used.

Test Questions

1. How do modern technologies impact cybersecurity?
2. How are the advanced technologies affecting cybersecurity?
3. Extensive data exposure: what are the risks?
4. Is there a strategy for controlling cyber breaches?
5. What is the reason for the shortage of cybersecurity professionals?
6. What impact do cyber-attacks have on businesses?
7. What are the main reasons for data exposure?
8. What is Risk?
9. How can cybersecurity affect national security?
10. What is Zero trust policy?

Bibliography

1. https://en.wikipedia.org/wiki/Cyber-physical_system
2. https://www.chalmers.se/en/areas-of-advance/ict/about%20us/Pages/default.aspx
3. https://www.bbvaresearch.com/wp-content/uploads/2017/09/maslow_pyramid_en.pdf
4. https://aginginplace.org/technology-in-our-life-today-and-how-it-has-changed/
5. https://www.sutori.com/en/story/history-of-ict-information-and-communications-techno logy--N7J51bQqSU7vLWcVfdn5M9qa
6. https://wiki.nus.edu.sg/display/cs1105groupreports/History+of+ICT
7. https://www.geeksforgeeks.org/difference-between-hardware-and-software/
8. https://www.educba.com/types-of-network-devices/
9. http://recherche.ircam.fr/anasyn/schwarz/da/specenv/3_1Digital_Signal_Processin.html
10. https://www.educba.com/types-of-network-devices/
11. https://www.microsoft.com/en-us/research/wp-content/uploads/2016/11/Input-Technolog ies-and-Techniques-HCI-Handbook-3rd-Edition.pdf
12. http://digitalthinkerhelp.com/output-devices-of-computer-types-examples-functions-uses/
13. https://www.techopedia.com/definition/3538/output-device
14. https://www.thecrazyprogrammer.com/2021/09/types-of-data-transmission.html
15. https://www.computerscience.gcse.guru/theory/storage-devices
16. https://en.wikipedia.org/wiki/Firmware
17. https://en.wikipedia.org/wiki/Operating_system
18. https://en.wikipedia.org/wiki/Communication_protocol#Protocol_design
19. https://en.wikipedia.org/wiki/Programming_language
20. https://www.slideshare.net/ebinrobinson/ict-importance-of-programming-and-progra mming-languages
21. https://www.uptech.team/blog/software-development-methodologies
22. https://www2.deloitte.com/us/en/insights/industry/public-sector/agile-in-government-by- the-numbers.html
23. https://www.complete-it.co.uk/the-history-of-information-technology/
24. https://www.geeksforgeeks.org/generations-of-computers-computer-fundamentals/
25. http://www.assis.pro.br/public_html/davereed/06-History.html
26. https://en.wikipedia.org/wiki/History_of_operating_systems
27. https://www.javatpoint.com/history-of-operating-system
28. http://www.rogerclarke.com/SOS/SwareGenns.html
29. https://www.laneways.agency/history-of-software-development/
30. https://en.wikipedia.org/wiki/A-0_System

K. Thakur et al., *Emerging ICT Technologies and Cybersecurity*,
https://doi.org/10.1007/978-3-031-27765-8

31. https://en.wikipedia.org/wiki/History_of_programming_languages
32. https://en.wikipedia.org/wiki/Programming_language_generations
33. https://www.rantecantennas.com/blog/the-different-types-of-wireless-communication/
34. https://www.ijert.org/generations-of-wireless-technology
35. https://en.wikipedia.org/wiki/5G
36. http://ijcsit.com/docs/Volume%205/vol5issue06/ijcsit20140506265.pdf
37. http://www.cbpp.uaa.alaska.edu/afef/web%20evolution%201-4.htm
38. https://www.frontierinternet.com/gateway/data-storage-timeline/
39. https://www.dataversity.net/brief-history-data-storage/#
40. https://en.wikipedia.org/wiki/DECtape
41. https://www.techopedia.com/definition/8210/magnetic-disk
42. https://www.encyclopedia.com/computing/dictionaries-thesauruses-pictures-and-press-rel
     eases/disk-cartridge
43. https://www.rohm.com/electronics-basics/memory/what-is-semiconductor-memory
44. https://www.electronics-notes.com/articles/electronic_components/semiconductor-ic-mem
     ory/memory-types-technologies.php
45. https://en.wikipedia.org/wiki/Flash_memory
46. https://www.britannica.com/technology/optical-storage
47. https://www.ibm.com/docs/en/i/7.1?topic=devices-optical-media-types
48. https://www.cdw.com/content/cdw/en/articles/datacenter/what-is-data-storage.html
49. https://www.red-gate.com/simple-talk/homepage/storage-101-data-center-storage-config
     urations/
50. https://www.datacore.com/software-defined-storage/
51. https://www.coursehero.com/file/p6jvadu/The-three-generations-of-software-development-
     are-defined-as-follows/
52. https://cloud.google.com/storage-transfer-service
53. https://en.wikipedia.org/wiki/Process_automation_system
54. https://www.dataversity.net/brief-history-analytics/
55. https://backendless.com/what-is-api-as-a-service/
56. https://techcrunch.com/2019/09/06/apis-are-the-next-big-saas-wave/
57. https://www.comptia.org/content/articles/what-is-cybersecurity
58. https://en.wikipedia.org/wiki/Digital_entertainment
59. https://medium.com/mckinsey-global-institute/next-generation-technologies-and-the-next-
     wave-of-globalization-376e6cfbcbb3
60. https://www.stateofai2019.com/chapter-2-why-is-ai-important/
61. https://en.wikipedia.org/wiki/Artificial_intelligence
62. https://www.britannica.com/technology/artificial-intelligence
63. https://www.ibm.com/cloud/learn/neural-networks
64. https://dr.lib.iastate.edu/server/api/core/bitstreams/a6de08e1-c74a-4235-8d92-ad7a92154
     733/content
65. https://becominghuman.ai/symbolic-ai-vs-connectionism-9f574d4f321f
66. https://www.elements-magazine.com/8-aims-and-objectives-of-artificial-intelligence/
67. https://www.geeksforgeeks.org/problem-solving-in-artificial-intelligence/
68. https://www.ibm.com/cloud/learn/natural-language-processing
69. https://jrodthoughts.medium.com/types-of-artificial-intelligence-learning-models-814e46
     eca30e
70. https://www.fingent.com/blog/classifying-knowledge-representation-in-artificial-intellige
     nce/
71. https://en.wikipedia.org/wiki/Machine_perception
72. https://www.javatpoint.com/history-of-artificial-intelligence
73. https://sitn.hms.harvard.edu/flash/2017/history-artificial-intelligence/
74. https://www.statista.com/statistics/607716/worldwide-artificial-intelligence-market-rev
     enues/
75. https://aibusiness.com/document.asp?doc_id=760184

76. https://www.simplilearn.com/tutorials/artificial-intelligence-tutorial/artificial-intelligence-applications
77. https://www.tutorialspoint.com/artificial_intelligence/artificial_intelligence_natural_language_processing.htm
78. https://byteiota.com/stages-of-nlp/
79. https://en.wikipedia.org/wiki/Computer_vision
80. https://en.wikipedia.org/wiki/Expert_system
81. https://www.britannica.com/technology/expert-system
82. https://en.wikipedia.org/wiki/Speech_recognition
83. https://aibusiness.com/author.asp?section_id=789&doc_id=773741
84. https://packagex.io/blog/technology/ocr-machine-learning/
85. https://www.techopedia.com/definition/9961/voice-recognition
86. https://www.readingrockets.org/article/text-speech-technology-what-it-and-how-it-works
87. https://www.freshworks.com/live-chat-software/chatbots/three-types-of-chatbots/
88. https://www.forbes.com/sites/cognitiveworld/2019/06/19/7-types-of-artificial-intelligence/?sh=c863bd5233ee
89. https://www.govtech.com/computing/understanding-the-four-types-of-artificial-intelligence.html
90. https://www.section.io/engineering-education/intelligent-agents-in-ai/
91. https://www.geeksforgeeks.org/types-of-environments-in-ai/
92. https://www.edureka.co/blog/artificial-intelligence-algorithms/
93. https://dzone.com/articles/exploring-ai-algorithms
94. https://machinelearningmastery.com/bayes-theorem-for-machine-learning/
95. https://monkeylearn.com/blog/introduction-to-support-vector-machines-svm/
96. https://www.javatpoint.com/k-nearest-neighbor-algorithm-for-machine-learning
97. https://machinelearningmastery.com/clustering-algorithms-with-python/
98. https://www.javatpoint.com/clustering-in-machine-learning
99. https://www.geeksforgeeks.org/ml-expectation-maximization-algorithm/
100. https://www.mygreatlearning.com/blog/introduction-to-multivariate-regression
101. https://www.techopedia.com/definition/33181/training-data
102. https://becominghuman.ai/what-is-training-data-its-types-and-why-it-is-important-f998424c3c9
103. https://www.telusinternational.com/solutions/ai-data-solutions/data-types
104. https://developer.qualcomm.com/software/qualcomm-neural-processing-sdk/learning-resources/ai-ml-android-neural-processing/data-collection-pre-processing
105. https://becominghuman.ai/what-is-the-difference-between-data-annotation-and-labeling-in-ai-ml-bc0cabe0b9d6
106. https://research.aimultiple.com/audio-annotation/
107. https://medium.com/@mnpinto/multi-label-audio-classification-7th-place-public-lb-solution-for-freesound-audio-tagging-2019-a7ccc0e0a02f
108. https://www.shaip.com/blog/guide-to-annotate-label-videos-for-ml/
109. https://kili-technology.com/blog/video-annotation-deep-learning
110. https://builtin.com/artificial-intelligence/artificial-intelligence-future
111. https://www.pewresearch.org/internet/2018/12/10/artificial-intelligence-and-the-future-of-humans/
112. https://mitsloan.mit.edu/ideas-made-to-matter/why-future-ai-future-work
113. https://www.statista.com/statistics/607716/worldwide-artificial-intelligence-market-revenues/
114. https://machinelearningmastery.com/gentle-introduction-gradient-boosting-algorithm-machine-learning/
115. https://mitsloan.mit.edu/ideas-made-to-matter/machine-learning-explained
116. https://www.ibm.com/cloud/learn/machine-learning
117. https://www2.deloitte.com/us/en/insights/focus/cognitive-technologies/state-of-ai-and-intelligent-automation-in-business-survey.html

118. https://www.fortunebusinessinsights.com/machine-learning-market-102226
119. https://www.simplilearn.com/tutorials/machine-learning-tutorial/what-is-machine-learning
120. https://www.ibm.com/cloud/learn/supervised-learning
121. https://www.ibm.com/cloud/learn/unsupervised-learning
122. https://www.javatpoint.com/semi-supervised-learning
123. https://www.geeksforgeeks.org/ml-semi-supervised-learning/
124. https://www.javatpoint.com/reinforcement-learning
125. https://www.geeksforgeeks.org/what-is-reinforcement-learning/
126. https://www.simplilearn.com/tutorials/deep-learning-tutorial/what-is-deep-learning
127. https://www.analyticssteps.com/blogs/top-6-machine-learning-techniques
128. https://machinelearningmastery.com/dimensionality-reduction-for-machine-learning/
129. https://www.simplilearn.com/tutorials/artificial-intelligence-tutorial/what-is-natural-lan
     guage-processing-nlp
130. https://www.geeksforgeeks.org/natural-language-processing-overview/
131. https://machinelearningmastery.com/what-are-word-embeddings/
132. https://www.javatpoint.com/applications-of-machine-learning
133. https://www.bmc.com/blogs/machine-learning-can-benefit-business/
134. https://www.researchandmarkets.com/reports/4806169/machine-learning-global-market-tra
     jectory-and
135. https://www.forbes.com/sites/forbestechcouncil/2019/10/30/12-impactful-ways-to-incorp
     orate-machine-learning-into-business-intelligence/?sh=5335baca6194
136. https://www.securityroundtable.org/the-growing-role-of-machine-learning-in-cybersecu
     rity/
137. https://www.computer.org/publications/tech-news/trends/the-impact-of-ai-on-cybersecurity
138. https://www.simplilearn.com/tutorials/blockchain-tutorial/blockchain-technology
139. https://www.ibm.com/topics/what-is-blockchain
140. https://www.forbes.com/advisor/investing/what-is-blockchain/
141. https://www.ndtv.com/business/the-first-bitcoin-transaction-was-for-buying-pizzas-more-
     interesting-tidbits-inside-2512643
142. https://www.blockchain-council.org/blockchain/cryptographic-hashing-a-complete-ove
     rview/
143. https://www.euromoney.com/learning/blockchain-explained/how-transactions-get-into-the-
     blockchain
144. https://www.investopedia.com/terms/p/proof-work.asp
145. https://www.investopedia.com/terms/b/block-bitcoin-block.asp
146. https://thetrustedweb.org/what-is-a-timestamp/
147. https://www.investopedia.com/terms/p/private-key.asp
148. https://builtin.com/blockchain
149. https://www.investopedia.com/terms/d/distributed-ledger-technology-dlt.asp
150. https://www.foley.com/en/insights/publications/2021/08/types-of-blockchain-public-pri
     vate-between
151. https://www.statista.com/statistics/1015362/worldwide-blockchain-technology-market-
     size/
152. https://www.researchandmarkets.com/reports/5546940/cryptocurrency-market-global-ind
     ustry-trends?
153. https://www.investopedia.com/non-fungible-tokens-nft-5115211
154. https://www.researchandmarkets.com/reports/5522321/non-fungible-token-global-market-
     report-2022-by?
155. https://www.ibm.com/topics/smart-contracts
156. https://www.verifiedmarketresearch.com/product/smart-contracts-market/
157. https://www.statista.com/statistics/1229290/blockchain-in-banking-and-financial-services-
     market-size/
158. https://hackernoon.com/which-countries-are-casting-voting-using-blockchain-s33j34ab
159. https://www.alliedmarketresearch.com/blockchain-supply-chain-market

160. https://www.ibm.com/blockchain/supply-chain
161. https://www.geeksforgeeks.org/role-of-blockchain-in-cybersecurity/
162. https://www.itbusinessedge.com/security/potential-use-cases-of-blockchain-technology-for-cybersecurity/
163. https://www.qualcomm.com/5g/what-is-5g
164. https://www.cisco.com/c/en/us/solutions/what-is-5g.html
165. https://www.pcmag.com/news/what-is-5g
166. https://cisomag.eccouncil.org/the-importance-of-5g-security-in-todays-world/
167. https://www.govtech.com/sponsored/why-we-need-5g-technology-and-what-it-means-for-society
168. https://oaji.net/articles/2017/1992-1515158039.pdf
169. https://www.thalesgroup.com/en/markets/digital-identity-and-security/mobile/inspired/5G
170. https://medium.com/@nagasanjayvijayan/5g-technology-an-overview-275cfb61cfd3
171. https://www.viavisolutions.com/en-us/5g-architecture
172. https://www.hindawi.com/journals/wcmc/2019/5264012/fig3/
173. https://www.3gpp.org/about-3gpp
174. https://en.wikipedia.org/wiki/5G_NR
175. https://www.electronics-notes.com/articles/connectivity/5g-mobile-wireless-cellular/5g-nr-new-radio.php
176. https://www.electronics-notes.com/articles/connectivity/5g-mobile-wireless-cellular/5g-ng-nextgen-core-network.php
177. https://en.wikipedia.org/wiki/LTE_Advanced
178. https://www.westbase.io/blog/what-is-lte-advanced-pro
179. https://www.brookings.edu/research/why-5g-requires-new-approaches-to-cybersecurity/
180. https://www.forbes.com/sites/forbestechcouncil/2021/10/29/why-5g-networks-are-disrupting-the-cybersecurity-industry/?sh=e9e8bee1fe9e
181. https://www.oracle.com/internet-of-things/what-is-iot/
182. https://www.interviewbit.com/blog/features-of-iot/
183. https://www.visionofhumanity.org/what-is-the-internet-of-things/
184. https://ethw.org/Internet_of_Things?
185. https://www.techslang.com/definition/what-is-ambient-intelligence/
186. https://www.controleng.com/articles/iot-to-ioat-internet-of-autonomous-things-devices-provides-solutions/
187. https://www.tutorialspoint.com/internet_of_things/internet_of_things_technology_and_protocols.htm
188. https://behrtech.com/blog/6-leading-types-of-iot-wireless-tech-and-their-best-use-cases/
189. https://itchronicles.com/iot/iot-network-architecture/
190. https://encyclopedia.pub/entry/8977
191. https://reports.valuates.com/market-reports/QYRE-Auto-26M10077/global-industrial-iot-iiot
192. https://seedscientific.com/how-much-data-is-created-every-day
193. https://www.techbusinessnews.com.au/the-impact-of-iot-on-cybersecurity/
194. https://www.ibm.com/docs/en/txseries/8.2?topic=overview-what-is-distributed-computing
195. https://www.vmware.com/topics/glossary/content/distributed-cloud.html
196. https://www.hpe.com/us/en/what-is/edge-to-cloud.html
197. https://chisw.com/blog/distributed-cloud-benefits/
198. https://www.tibco.com/reference-center/what-is-distributed-cloud-computing
199. https://www.spiceworks.com/tech/cloud/articles/what-is-distributed-computing/
200. https://marketdigits.com/distributed-cloud-market/
201. https://en.wikipedia.org/wiki/Content_delivery_network
202. https://www.spiceworks.com/tech/networking/articles/what-is-content-delivery-network/
203. https://www.techopedia.com/definition/33000/software-defined-infrastructure-sdi
204. https://www.sciencedirect.com/topics/computer-science/big-data-processing

205. https://www.ibm.com/cloud/blog/distributed-cloud-vs-hybrid-cloud-vs-multicloud-vs-edge-computing-part-2
206. https://www.windzr.com/blog/the-challenge-of-distributed-cloud-architecture
207. https://www.f5.com/company/blog/redefining-cybersecurity-at-the-distributed-cloud-edge-with-ai-and-real-time-telemetry
208. https://www.ibm.com/topics/quantum-computing
209. https://en.wikipedia.org/wiki/Quantum_computing
210. https://www.investopedia.com/terms/q/quantum-computing.asp
211. https://docs.microsoft.com/en-us/azure/quantum/concepts-overview
212. https://www.britannica.com/technology/quantum-computer
213. https://deepai.org/machine-learning-glossary-and-terms/quantum-computation-theory
214. https://www.allaboutcircuits.com/technical-articles/fundamentals-of-quantum-computing/
215. https://cosmosmagazine.com/science/quantum-computing-for-the-qubit-curious/
216. https://www.quantum-inspire.com/kbase/superposition-and-entanglement/
217. https://www.cbinsights.com/research/quantum-computing-classical-computing-comparison-infographic/
218. https://hbr.org/2021/07/quantum-computing-is-coming-what-can-it-do
219. https://www.idquantique.com/quantum-safe-security/quantum-computing/real-world-applications/
220. https://research.ibm.com/blog/ibm-quantum-roadmap-2025
221. https://www.honeywell.com/us/en/company/quantum
222. https://quantumai.google/learn/map
223. https://www.theverge.com/2019/10/23/20928294/google-quantum-supremacy-sycamore-computer-qubit-milestone
224. https://news.microsoft.com/innovation-stories/azure-quantum-majorana-topological-qubit/
225. https://azure.microsoft.com/en-us/solutions/quantum-computing/#quantum-impact
226. https://www.energy.gov/science/articles/creating-heart-quantum-computer-developing-qubits
227. http://ffden-2.phys.uaf.edu/113.web.stuff/travis/what_is.html
228. https://winnerscience.com/type-i-and-type-ii-superconductors/
229. https://sciencenotes.org/superfluidity-definition-and-examples/
230. https://www.livescience.com/33816-quantum-mechanics-explanation.html
231. https://azure.microsoft.com/en-us/resources/cloud-computing-dictionary/what-is-a-qubit/#qubit-vs-bit
232. https://en.wikipedia.org/wiki/Quantum_logic_gate
233. https://en.wikipedia.org/wiki/Quantum_counting_algorithm
234. https://towardsdatascience.com/grovers-search-algorithm-simplified-4d4266bae29e
235. https://en.wikipedia.org/wiki/Shor%27s_algorithm
236. https://www.scientificamerican.com/article/what-are-josephson-juncti/#
237. https://www.cnet.com/tech/computing/vr-and-ar-looked-to-the-metaverse-at-ces-2022/
238. https://nextbridge.com/tactile-virtual-reality-technology/
239. https://en.wikipedia.org/wiki/Augmented_reality
240. https://www.vrs.org.uk/virtual-reality/history.html
241. https://heizenrader.com/the-3-types-of-virtual-reality/
242. https://www.physio-pedia.com/Somatosensation
243. https://www.ncbi.nlm.nih.gov/pmc/articles/PMC3172606/
244. https://www.gartner.com/en/information-technology/glossary/head-mounted-displays-hmd
245. https://www.sciencedirect.com/science/article/abs/pii/S0306457318310811
246. https://www.sciencedirect.com/science/article/pii/S2096579619300063
247. https://www.techtarget.com/whatis/definition/simulator-sickness
248. https://www.fortunebusinessinsights.com/industry-reports/virtual-reality-market-101378
249. https://www.techtarget.com/whatis/definition/virtual-reality-gaming-VR-gaming
250. https://www.fortunebusinessinsights.com/industry-reports/virtual-reality-gaming-market-100271

251. https://www.thebusinessresearchcompany.com/report/virtual-reality-in-education-global-market-report
252. https://www.analyticssteps.com/blogs/5-applications-virtual-reality-education
253. https://www.forbes.com/sites/forbesbusinesscouncil/2020/06/12/virtual-reality-a-game-changer-for-product-development/?sh=6837a6c96f99
254. https://www.futurebusinesstech.com/blog/the-world-in-2050-top-20-future-technologies
255. https://en.wikipedia.org/wiki/3D_printing
256. Ahmed, A., Arya, S., Gupta, V., Furukawa, H., and Khosla, A., "4D printing: Fundamentals, materials, applications and challenges," Polymer, Volume 228, Elsevier, 2021, 123,926.
257. https://www.marketsandmarkets.com/Market-Reports/4d-printing-market-3084180.html
258. https://en.wikipedia.org/wiki/6G_(network)
259. https://www.techtarget.com/searchnetworking/definition/6G
260. https://en.wikipedia.org/wiki/Autonomous_robot
261. https://locusrobotics.com/what-are-autonomous-robots/
262. https://www.techtarget.com/searchcio/definition/artificial-neuron
263. https://computerhistory.org/blog/how-do-neural-network-systems-work/
264. https://www.spiceworks.com/tech/artificial-intelligence/articles/narrow-general-super-ai-difference/
265. https://www.techtarget.com/searchenterpriseai/definition/artificial-general-intellige nce-AGI
266. https://www.techtarget.com/searchenterpriseai/definition/artificial-superintelligence-ASI
267. https://en.wikipedia.org/wiki/Mind_uploading
268. https://www.livescience.com/37499-immortality-by-2045-conference.html
269. https://en.wikipedia.org/wiki/Self-driving_car
270. https://www.aabb.org/news-resources/resources/cellular-therapies/facts-about-cellular-therapies/regenerative-medicine
271. https://en.wikipedia.org/wiki/Digital_twin
272. https://www.docseducation.com/blog/living-robots-poised-advance-modern-medicine
273. https://www.npr.org/2021/12/01/1060027395/robots-xenobots-living-self-replicating-copy
274. https://www.airswift.com/blog/human-augmentation
275. https://onepullwire.com/news/types-industrial-automated-systems/
276. https://www.advsyscon.com/blog/what-is-intelligent-process-automation/
277. https://en.wikipedia.org/wiki/Space_elevator
278. https://en.wikipedia.org/wiki/Skyhook_(cable)
279. https://www.planetary.org/articles/what-is-solar-sailing
280. https://techreport.com/blog/3476302/evolving-technology-cybersecurity/
281. https://www.cyberdegrees.org/resources/hot-technologies-cyber-security/
282. https://cio.economictimes.indiatimes.com/news/digital-security/disruptive-impact-of-emerging-technologies-on-cyber-security/87922311
283. https://www.strategicmarketresearch.com/market-report/cyber-security-market
284. https://cybersecurityventures.com/cybercrime-damages-6-trillion-by-2021/
285. https://ccdcoe.org/uploads/2018/10/Hare-The-Cyber-Threat-to-National-Security-Why-Cant-We-Agree.pdf
286. https://www.statista.com/statistics/273550/data-breaches-recorded-in-the-united-states-by-number-of-breaches-and-records-exposed/
287. https://cybersecurityventures.com/cybersecurity-spending-2021-2025/
288. https://www.livemint.com/technology/shortage-of-cybersecurity-professionals-a-key-worry-for-firms-in-22-11642015098080.html